POWERING THE DREAM

POWERING THE DREAM

The History and Promise of Green Technology

ALEXIS MADRIGAL

DA CAPO PRESS

A Member of the Perseus Books Group

Editorial production by the Book Factory.
Designed and set in 11 point Minion Pro by Cynthia Young at Sagecraft.

**Cataloging-in-Publication data for this book is available from the
Library of Congress.**
ISBN: 978-0-306-81885-1

Published by Da Capo Press
A Member of the Perseus Books Group
www.dacapopress.com

Da Capo Press books are available at special discounts for bulk purchases in the
U.S. by corporations, institutions, and other organizations. For more information,
please contact the Special Markets Department at the Perseus Books Group,
2300 Chestnut Street, Suite 200, Philadelphia, PA 19103, or call (800) 810-4145,
ext. 5000, or e-mail special.markets@perseusbooks.com.

10 9 8 7 6 5 4 3 2 1

To S. L. M. and E. K. M.,
who know who they are.

CONTENTS

PREFACE

IN THE AMERICAN DREAM, everyone can own a home and a couple of cars, and we can go wherever we damn well please. There are good jobs for hard workers and open roads for the weekends. We can move both up and around.

All of these things seem enshrined in our interpretation of the Constitution: Yes, this is what it means to have life, liberty, property, and the pursuit of happiness. But really what allows us to have the classic American life is a material constitution, the steel and concrete and energy that power our country's great machine. This constitution, as life shaping and influential as the one penned by the founding fathers, is what this book addresses.

We're going to look at solar and wind machines, ways of building houses and developments, failures and successes. The point isn't to find old solar machines we can use again but instead to understand how our country got built. Just as understanding the Constitution is difficult without knowing something about other nations' founding documents, we can't understand the choices we've made without understanding how people in the past saw the technological playing field. The detours and off-ramps of our history are important as a record of choices not made, as the shadows of the stars of history. Without them, there's no "honest" record of how our current infrastructure and technical reality came to be. And without that, figuring out why we have the world we do is hard. Old technologies tell us a lot about how society developed: what forces drove what, who benefited, what was gained, and what was lost.

No other group of people has been more enamored of power and the technology it undergirds. Appreciation for the technological sublime—from the skyscrapers of Manhattan to the rocket launches of Apollo—is not unique to Americans, but boy, do we do it well. For better and for worse.

We Americans have sworn always to be warm, no matter what temperatures provided, and always to see, no matter how much light was reaching us from our local star. We have adopted the machines of energy conversion—automobiles, power plants, trains, water wheels, windmills—with a ferocity that simultaneously frightens and entrances rich and poor countries alike. With increasing speed after the Civil War, we force-fed our economy with immigrants and energy stored in the form of coal, then oil. Cities expanded and grew increasingly networked together. Energy infrastructure extended from sea to shining sea.

Americans can go anywhere they want in cheap cars that move as if pulled by three hundred horses. We use a kilowatt-hour of electricity—equivalent to something like fifteen husky gentlemen's maximum muscle power—for a few cents. There's not a person in America who doesn't benefit from living in an energy-besotted nation.

But the impacts of the energy sources we use have come up against other things that we care about, like the health of our children, the preservation of other kinds of life, and our national credibility. All kinds of energy production have negative impacts, but because they're all different, picking out which ones are compatible with our other ideals can seem difficult.

The pugilists in the energy wars think they can win by proving their technology is the most inevitable inevitability. We have to go solar, they say. We have to go nuclear, others say. We need to keep burning coal. We need more drilling. We need to use less oil. Growth is the answer. Growth is the problem.

But there will be no magic bullet. We could destroy the things we love. Technology can be, but is not always, the answer. Ideas about nature matter. Attitudes and mistakes and misapprehensions are as much a part of energy history as the heat content of coal. So this book is a book of stories about this country, submitted with hard-won humility I inculcated by seeing how wrong people have been about energy down through the decades.

I selected the microhistories in this book for their connection to the present, not their importance to contemporaries. With a hot field like green technology, the tendency is to become very future focused. Innovators rush into the field to build new stuff—a change is gonna come!

But people do not always succeed in executing even the best-laid plans: Random events have major impacts and people take bad paths. However, simply remembering what actually happened—the sense of historicity (or whatever you might call it)—is what I hope forms the lasting value of this book. If history's unique contribution is that it can help us understand how imperfect our information is and how viscous cultural norms and physical realities place limits on action, then green technology is a field in desperate need of some historical perspective.

Delve deep enough into events that seem obvious now, like electricity's win over compressed air as a way of transmitting power over long distances, and you find counterarguments and uncertainty. Likewise, trends that seemed bulletproof during the 1970s energy shocks, like the gradual decrease of large, gas-guzzling cars, were riddled with (unleaded) holes by the rise of the sport utility vehicle.

Conspiracy theories and triumphant narratives seem to make sense in the face of such tremendous complexity: It's the corporations' fault—all of it—or the greens killed off nuclear power! Oil sheikhs, utility executives, General Motors, and communist-sympathizing hippies all figure prominently in the mythmaking that surrounds our energy past and present. However, I have rarely found the big movements in energy to hinge on such characters, despite my best attempts to find their fingerprints in my investigations.

This book is about the uncertainties and triumphs of innovation, the mysterious process by which ideas are made into products out there in the world. In a realm in which everyone argues that things must or will happen, the knowledge of our fallibility is what is most important. It's not in the strengths of our arguments where we can find common ground with others but rather in their weaknesses. Perhaps the less sure we are of the directions we're taking, the more carefully we'll step and the more often we'll stop to ask for directions.

ACKNOWLEDGMENTS

INNUMERABLE PEOPLE CONTRIBUTED to the making of this book. Writing may seem like a solitary pursuit, but you can only do it well when a bunch of other people band together to make the space for you to do your work. Beyond all others, Sarah Rich will always have my love and thanks for the way she made space and time for me by running our lives during the long months of typing. She was the first to believe in me, the closest to me when my brain was stationed in the past, and the last reader before the printer. There are not enough letterpress thank-you cards in the world.

The University of California–Berkeley's Office for the History of Technology gave me a workspace, both literally and figuratively. I owe much of my success in exploring new territory here to their generosity.

I also have the deepest gratitude to *Wired* and the *Atlantic,* which provided wonderful institutional support for this endeavor. Specifically, my old boss and friend Betsy Mason deserves a million high fives. She taught me much about everything and always indulged (and secretly shared) my passion for weird history. Thank you also to my new boss, Bob Cohn, for his unflagging support. He is the best manager in the magazine business, according to everyone who has ever worked with him, including me.

On the business end of things, I could not be happier with my agent David Fugate, who helped me shape a bunch of ideas into something with purpose, and Bob Pigeon, my editor and the kind of guy who always knows just what to say.

Many people deepened my thinking and challenged my assumptions, including Myra Rich, Frank Laird, Andrew Kirk, Adam Rome, Peter Shulman, Erle Ellis, Chris Mims, Andy Revkin, Gregor MacDonald, Michael Kanellos, John Pavlus, Chris Nelder, Brandon Keim, Dave Roberts, Todd Woody, Jesse Jenkins, Brian Walsh, Seth Garz, Heather Fleming, Randy Alfred, Ruth Seeley, Robin Sloan, Matthew Battles,

Tim Maly, Geoff Manaugh, Nicola Twilley, Maggie Koerth-Baker, Sascha Pohflepp, Tim Carmody, Maureen Ogle, and Robert Rich.

And to those who kept me sane in darkest days, I owe you several million drinks each: my parents, Marissa Madrigal, Alex Miel, Teddy Wright, Ziggy Whitman, Jessica Battilana, Sarah Picard, Cara deFabio, Tali Horowitz, Noah Rauch, Joel Snyder, Betsy Uhrman, Julie Sherwood, Dylan Fareed, Yaron Milgrom-Elcott, Miriam Sheinbein, Mat Honan, Harper Honan, Kristin Smith, and Jon Snyder.

Though they don't know it, the pioneers of deep thinking about energy and society—living and dead—are who kept my mind alive in writing this book. So to Stewart Brand, J. Baldwin, Palmer Putnam, Alvin Weinberg, Arnold Goldman, and many others, thank you for your foresight.

I should also thank Ritual Coffee, Coffee Bar, and that weird place by the 24th Street BART Stop in San Francisco for allowing me to take up valuable table space. Couldn't have done it without your coffee or hospitality. And of course, I should thank Twitter, which makes the long days seem shorter.

Introduction

Y OU COULD BE EXCUSED for thinking that there is no history of what we call green technology. If you're reading *Time* or watching television or listening to Sarah Palin or Barack Obama, according to them, solar and wind are new, geothermal has barely been tried, and efficiency begins tomorrow. Perhaps some hippies toyed with off-grid living in the '60s and there was some boondoggle in the Carter administration, but the hippies abandoned their communes in the woods and Carter lost to Reagan, so nothing much happened. Maybe a few windmills once creaked on your great grandfather's farm, or you've heard him tell about how the Tennessee Valley Authority dammed a river with some farm boys ("Just hear their hammers ringing / They'll build that dam or bust."). General Motors killed the electric car, too, right?

According to those who do remember a few facts from the past, either we should be whizzing around gleaming clean cities in silent cars powered solely by solar panels and micro-turbines or green tech is an engineering bust that has been given ample opportunity to prosper, but it has only succeeded in failing spectacularly.

There's almost no institutional memory of what happened before the energy crises of the '70s, and little of what happened technologically during that time has documented in any serious way. Far more people know about the Enola Gay or the rise of Disneyland or the demise of the spotted owl than about any solar, wind, wave, water, or geothermal project.

Nonetheless, some remnants of the past endure. In 2007, when I typed the word "solar" into the search box of the *American Memory* collection on the Library of Congress Web site, I got one good search result. The link I clicked read, "Death Valley Ranch, Solar Heater, Death Valley Junction vicinity, Inyo county, CA." Three black-and-white photos and an architectural drawing provided details.

They show a rotting wooden building on a concrete foundation maybe sixty feet long and nine feet wide, taller in the back than the front, so the front surface slopes at a 36-degree angle. It's covered in copper metal coils that are painted black and snake back and forth. Behind it a tall cylinder, wrapped in felt made from cattle hair and what looks like aluminum foil, rises twenty feet into the air. The entire scene is surrounded by desert—cactus, rock, sand, sky. The caption reads, "The Solar Heater at Death Valley Ranch is a rare surviving example of a solar industry that thrived in Southern California before World War II and before the widespread use of natural gas."[1]

Wait, what?

There was a flourishing solar industry in California before World War II? Why did people start using solar heaters? If they worked, why did they stop? (You can read all about it on pages 84–89.)

As my research continued, I found the six million windmills of the prairies, the California wave motor craze of the 1890s, the electric cars of the early 1900s, the solar home boom of mid-century, the world's first megawatt wind turbine, which went online in 1941, the oil companies' contribution to photovoltaics, decades-old algal biodiesel programs, and the huge solar farm of the Reagan years.

The history was long and deep, but criminally obscure. It's understandable; victors don't only write the history in military battles. The popular view of technology is that the best one wins. We assume that alternatives did not exist or that, if they did, they were obviously and irreversibly inferior to the options that were chosen.

Recent historians of technology have pounded away at this way of thinking. One of the best, Imperial College of London's David Edgerton, has a simple remedy for fixing this cognitive blind spot: Forget calling all these human-made objects and systems technology and call them "things" instead. "Thinking about the use of things, rather than of technology, connects us directly with the world we know rather than the strange world in which 'technology' lives," Edgerton wrote in his superb history, *The Shock of the Old*.[2]

Furthermore, things don't have to be radically better to beat out other things. Many technologies persist, even if they aren't dominant. "The paper-clip is ubiquitous not because it is an earth-shatteringly important technology," Edgerton pointed out. "There are many ways of holding paper together: pin it, staple it, punch holes and secure it with 'Treasury tags,' use Sellotape, put it in a ring-bind or other sort of folder, or bind it into a book. We use paper-clips so much because they are, for many uses, marginally better than the alternatives, and we know this."[3]

Another example we're all familiar with is the ubiquity of Microsoft Word. There are dozens of ways to do word processing. During PCs' early march into businesses and homes, there were all kinds of software programs for typing, formatting, saving, sending, and printing documents. They all pretty much worked.

Then, Microsoft Office—for a variety of business, not technical reasons—began to gain market share.[4] For example, it often came pre-installed on new PCs, which already ran Microsoft-made operating systems. Microsoft competed fiercely and won, but few believe "the best technology" won. Word worked, and Microsoft was a strong force in the market, but it wasn't as if there were no alternatives.[5] Microsoft Word and Excel became the de facto standard for exchanging files between users, thereby also increasing the momentum of their product. Using Microsoft Word simply became easier than completing the series of tiny actions necessary to convert some other word processor's native output into a file that Microsoft Word could read.

Economists describe this process as "the network effect," and it's associated with all kinds of positive feedbacks as success breeds success.[6] Historians sometimes call the broader version of this phenomenon "technological momentum."[7] Its effect is to transform a series of marginal choices ("What the hell, maybe I will use Word, not WordPerfect") into market dominance.

If we run a lot of energy technologies through the paper clip and Microsoft Word filters, we find many of the same patterns. Because energy technologies tend to require a lot of concrete and steel and money, momentum is even more important. A decision made at a particular moment for particular reasons will have repercussions for decades; after all, the Hoover Dam is now almost seventy-five years old.

That's why the solar hot water heater and the rest of the projects in this book aren't mere curiosities. They got researched and built for a reason. Some people, at some time, thought they were a viable alternative to the systems in place. We have to go back to that moment when someone picked Word or a gasoline-powered car and ask, Why was that choice made?

Political scientist Langdon Winner faulted a predecessor of green technology, the "small is beautiful" school of appropriate technology, for a "grievous" historical amnesia. "Those active in the field," he wrote, "were willing to proceed as if history and existing institutional technical realities did not matter." This book is an attempt to answer the questions that Winner thinks an insurgent technological group has to address if they want to make their revolution:

> One ought to be able to discover points at which developments in a given field took an unfortunate turn, points at which the choices produced an undesirable instrumental regime. One could, for example, survey the range of discoveries, inventions, industries, and large-scale systems that have arisen during the past century and notice which paths in modern technology have been selected. One might then attempt to answer such questions as, Why did the developments proceed as they did? Were there any real alternatives? Why weren't those alternatives selected at the time? How could any such alternatives be reclaimed now?[8]

The next section delves into the alternative energy technologies that existed over the last century. We'll find good ideas that were dropped and bad ideas that were probably better forgotten. We'll see mad inventors trying to navigate the choppy seas of fossil fuel prices, bad luck, dirty dealing, a lack of government support, societal shifts, competing new energy technologies, and a host of other factors. Despite being tossed about, many of them succeeded in creating real alternatives that merely lacked funding and scale, not technical sophistication. The bottom line is that we've missed chances to have a cleaner energy system, and if we don't heed the lessons of the past, we could blow this opportunity, too.

In 1900 people could use the sun to heat the water for a shower. They could drive across New York City in an electric taxicab. Even if these cabs did not work perfectly, they existed before most people even had a single light bulb in their home. In 1945 a person could have purchased a solar house or gone to see the one-megawatt wind turbine. During the 1970s one could have visited the Solar Energy Research Institute and, ten years later, seen the massive solar fields of the Mojave Desert. Green technology has been a viable set of technologies for more than one hundred years but, regardless, supplies little of America's energy. What happened? What might have been?

Let's find out.

The Dream of a More Perfect Power I.

chapter 1

Profit, Salvation

E VERY YEAR, THE WORLD'S TECHNO-ELITE gather at the TED Conference in California to talk about the future of every-thing. The event's speakers tend to focus on technologies and ideas that are, as they put it, "worth spreading." They aim to influence. The audience is wealthy and smart, so a good idea presented well can catapult to a very powerful kind of niche fame. So, when famed venture capitalist John Doerr stepped onto the stage in 2007 and cried about global warming, it signaled that something very strange was about to happen to environmentalism in America. "I'm really scared," he began, waving an index card. "I don't think we're going to make it."

Over the next seventeen minutes, Doerr revealed his own come-to-Jesus moment and heralded the arrival of a potent new pro-technology force in environmentalism. He recounted a dinner he held for a large group, at which the discussion turned to climate change. One by one, the party attendees gave lip service to the problem. Then, it was his daughter's turn. "Dad, your generation created this problem," he described the teenager saying. "You better fix it." And just like that, Doerr says, "everything changed."

He and his partners at Kleiner Perkins Caufield & Byers, made rich and famous from early investments in Google and Amazon, fanned out across the globe looking for solutions to "fix" global warming. "We may have the political will to do this in the US, but I gotta tell you, we have only one atmosphere, so somehow we're going to have to find the political will to do this all around the world," Doerr said. "We've got to make this economic so that all people and all nations make the right outcome the profitable outcome and therefore the likely outcome."

The machinery that environmentalists had used to secure clean air and water—policies, limits, and laws—would be necessary but not sufficient. Doerr's solution for climate change would require innovation and infrastructure building on a massive scale; stopping power plants from being built would not be enough. New ones—clean ones—would have to be built.

That such a rollout could occur seems possible, even plausible, to Doerr. He and many of those in the crowd had used personal computer and networking technologies to reshape the world. "I can't wait to see what we TEDsters do about this crisis," he said, "and I really, really hope that we multiply all of our energy, all of our talent, and all of our influence to solve this problem." Choking up, he retreated to a chair, the lone prop on the stage. "Because if we do," he paused, verklempt again, then concluded, "I can look forward to the conversation I'm going to have with my daughter in twenty years."

The powerful talk ended as abruptly as it began. On his way off the stage, a man approached and bear hugged him. The clapping, which had begun slowly, grew louder as the crowd realized the presentation was over. People began to stand. Seemingly oblivious to the reception, Doerr walked off the stage into the arms of a woman whom he gave a fierce hug, his face contorting with obvious emotion as the audience settled into the resonant frequency of a standing ovation.[1]

It was, in the words of fellow venture capitalist and pundit Paul Kedrosky, "one of the strangest moments . . . at an admittedly often strange conference."[2] Either Doerr was a remarkably good actor—certainly a plausible scenario—or he was genuinely, deeply worried about the state of the world.

Perhaps an environmentalist crying over the state of the world would not exactly count as hot news. But John Doerr started his career at Intel and holds master's degrees in electrical engineering and business, not ecology or political science. He and the venture capital firm he worked for, Kleiner Perkins Caufield & Byers, were the very people who refused to see limits to growth. He was not the guy one expects to cry about the earth on the tech elite's biggest stage. It was a sign that a new type of movement to save the world was on. In response to the threat of global warming and the opportunity to take down the Jurassic energy industry, the green banner was pulled from the musty closet of a previous generation's lexicon and unfurled on a new streamlined boat. Green tech was born.

The prime actors in the green tech movement aren't activists. Few would advocate protests as a means to solving problems. Instead, their mantra is creative destruction, and their targets are the wildcatters, utilities, power plant makers, and infrastructure maintainers that keep the country's grid humming and its cars filled with fuel.

Drawing on their experience watching the price of computing power inexorably fall along the path termed Moore's Law, after the Intel chief who called out the trend, the green technologists are certainly something new on the scene. Their faith that the methods of technology can solve energy problems may be a virtue, or their optimism may be, in Vaclav Smil's evocative language, "Moore's Curse."[3] For now, we are able to say for certain that Silicon Valley, with its chapbook of quirks, entered the energy fray in the first decade of the twenty-first century. In 2008 and 2009 venture capital firms doled out $12.5 billion to green-tech startups,[4] and they've already exerted a powerful influence on the nation's leaders' thinking about energy and environmental issues. In fact, John Doerr is now a special economic adviser to the Obama administration.[5] Furthermore, in early 2010 Bill Gates gave his own TED talk, calling for low-carbon "energy miracles."[6]

The technological base the green technologists are working from has surprisingly deep roots. The basics of the grid-and-oil energy system rounded into place a little more than a century ago. As we'll see, wave motors, wind turbines, solar power plants, electric cars, and a host of other "alternative" energy ideas had already seen the light of day long before the average American had a single light bulb to fend off the darkness. If we don't know these histories, it's because the fossil-fueled economy of the twentieth century had a tendency to pave over alternatives to itself, leaving only curious hints of worlds that might have been.

But Doerr and his venture capital cohorts, such as Ira Ehrenpreis of Technology Partners and Erik Straser of Mohr Davidow Ventures, don't see the developments of the last century as a bad thing. The globalization of markets, the expansions of economies, the triumph of capitalism over religious, ethnic, and societal resistance—these are what made the material world we all inhabit. Like most techno-optimists, they believe it's a much better world than that of 1957 or 1907 or 1807. "I'm a raging capitalist. My job is to make a lot of money," Doerr told a reporter in 2009. "I'm a technology junkie. I'm also an American. I'm a very lucky kid from St. Louis."[7]

Doerr doesn't want to unmake the industrial world—with all its inequities and problems; rather, he wants to remake it, to sustain it, to grow it. And to make money from it: "Energy is a six trillion dollar business worldwide. It is the mother of all markets," Doerr told the TED crowd. "Remember that Internet? I'll tell you what: Green technology—going green—is bigger than the Internet. It could be the biggest economic opportunity of the twenty-first century."

Doerr's talk is listed on the TED website under the heading, "John Doerr Sees Salvation and Profit in Greentech." To the intellectual environmentalists of the 1960s and '70s, however, Doerr's notions of nationalism and progress—with a capital P—would have seemed quaint. Herman Daly, a radical economist, who was very prominent in the alternative energy circles of the early 1970s, attacked the very notions of economic growth. "As [another economist] so bluntly put it in defending growth: 'Growth is a substitute for equality of income. So long as there is growth there is hope, and that makes large income differentials tolerable,'" Daly told a Senate committee in 1973. "We are addicted to growth because we are addicted to large inequalities in income and wealth. What about the poor? Let them eat growth! Better yet, let them feed on the hope of eating growth in the future."[8]

Daly ended up being quoted by Amory Lovins, Hazel Henderson, and a host of other less prominent environmental sources. Even in the late 1990s "fourth-wave environmentalist" Leslie Paul Thiele could write in his book *Environmentalism for a New Millennium*, "Environmentalists often find themselves isolated in their reluctance to join the popular celebration of economic growth."[9]

Thus, although the new green technologists don't fit well into the standard stories of the environmental movement, this book outlines a different set of renewable energy entrepreneurs who were driven by goals much closer to Doerr's than Daly's. In fact, perhaps the first green futurist, John Etzler, imagined the machines and society that could form "the utopian origins of economic growth."[10]

chapter 2

The First Green-Technology Futurist

JOHN ETZLER WAS PROBABLY CRAZY, but not so much more than your average futurist. A friend of John Roebling, who built the Brooklyn Bridge, Etzler weaves in and out of history in the first half of the nineteenth century, showing up in odd locales and never quite being in the right place at the right time. We know him best for a slim, cockeyed volume he wrote in the 1830s called *The Paradise Within the Reach of All Men, Without Labour, by Powers of Nature and Machinery: An Address to All Intelligent Men*. It came with two afterwords addressed to President Andrew Jackson and Congress. Filled with mental sketches for tremendous machines that would harness the wind, waves, and sun for human purposes, the book rests on an insight blooming with Industrial Revolutionary fervor that has been repeated a million times over by solar advocates:[1]

The substance of this book is—
1. It is proved that there are powers at the disposal of man, million times greater than all human exertions could effect hitherto. These powers are derived—
   a. From wind
   b. From the tide
   c. From the waves of the sea, caused by wind
   d. From steam, generated by heat of the sun, by means of concentrating reflectors or burning mirrors of simple contrivance

The sun and wind and waves represented an infinite power. With such incredible energy stores, the sky was, literally, the limit for Etzler's imagination. Written at a time when oil hadn't ever powered an engine and coal use was concentrated mostly in England, his rabid enthusiasm led him to imagine, futuristlike, many components of the high-energy society we live in: plastic-based product culture, industrialized food manufacture, apartments with elevators and air conditioning, synthetic fibers, and huge vehicles that would not need rails to go forty miles per hour. Sounds familiar, right?

"Etzler designed not a world to come, but the world that came," the historian Steven Stoll concluded. Stoll locates Etzler's peculiar prescience in "his sense that human happiness would be understood as the application of technology to convenience and leisure."[2]

With the right energy sources, *of course* we'd have anything an able mind or a marketing executive could conjure!

Etzler's thinking was heavily influenced by Georg Wilhelm Friedrich Hegel, the preeminent German philosopher of the day. Hegel believed human history had an arrow, that it was going someplace, that it was progressive. Even better, the world, as directed by the human mind, was slowly being perfected. Karl Marx and Friedrich Engels would then turn this idea into a radical call for revolution. Earlier social reformers, however, usually had smaller dreams: They just wanted better communities that might take some of the edge off the shock of industrialization.

Transplanted to the American soil, these ideas took the form of utopian adventures. Dozens of them sprung up all over the nation, but particularly in what was then thought of as the West: Ohio, Kentucky, and Indiana.[3] While wages were rising,[4] disturbing things were happening to the people who were paid that money. The average life expectancy of even native-born white American males started to drop around 1800 and did not recover fully until the 1940s. Beginning around 1830 children were, on average, a little shorter than their predecessors, a sign that the extra cash in their pockets was not improving their health and well-being.[5] Incredible dislocations were beginning to occur in American society, and many could see that these changes were just the start of a much greater movement.

Anxiously looking over at the industrial towns of Great Britain, Americans saw coal-burning centers like Manchester and London shrouded with a permanent cloud of smoke and soot.[6] The working classes of the cities were unhealthy and dying, with their children toiling in factories. Quantitative standard-of-living measures aside, the unique horrors of the British cities of the early industrial period were well known.[7] The factories were the world of Oliver Twist and his trembling voice, asking, "Please, sir, I want some more."[8]

The American utopias were direct antecedents for Etzler's ideas. In the year he wrote his book, Etzler spent time at the German-Christian utopianist George Rapp's community "Economy." He also hung around New Harmony, Indiana, a thirty thousand–acre township that Rapp had sold to the wealthy British factory reformer Robert Owen. Owen's experiment, like many others of the time, failed miserably, devolving into petty squabbles in only a few years. Many of these early socialist communes were detailed in 1870 by John Humphrey Noyes, himself a founder of an oddly successful utopia with an open-marriage policy.

Noyes, like Etzler, found the communes' insistence on working the land both boring and wrong-headed.[9] They were back-to-the-landers, not manufacturers. "Almost any kind of a factory would be better than a farm for a Community nursery," he complained. He mocked the lack of technology on the communes of the day, lamenting that "the saw-mill is the only form of mechanism" often seen. "It is really ludicrous to see how uniformly an old saw-mill turns up in connection with each Association, and how zealously the brethren made much of it; but that is about all they attempted in the line of manufacturing. Land, land, land, was evidently regarded by them as the mother of all gain and comfort," Noyes complained.[10]

Etzler agreed. Surveying the social utopian scene of the 1830s, Etzler found something lacking: proper attention to the role of energy. If they wanted to change the world, he contended, the communitarians would need power—and lots of it. Coal did not appeal to him as an energy source, as he saw "industrialism as a vicious energy monopoly. Nothing but the cost of coal dictated that the Many would sweat for wages in factories owned by the Few."[11] Renewable energy, however, fell everywhere on everyone. It was an unlimited, democratic source of power.

In the way that Marx (another Hegelian) believed the proletariat would overthrow the bourgeoisie and remnants of aristocracy, Etzler believed that this energy source would displace fossil fuels and human power. All that was needed was to change the technology that humans used to power their civilization. In other words, change the energy system and we could change the way that men related to each other. Evil would ebb and perfection would rush into the society.

So before industrialization had fully washed up on the shores of America, Etzler had hit upon the pursuit of a more perfect power. The wind and the sun and the waves, Etzler wrote, "are more than sufficient to produce a total revolution of the human race." All of America could be "changed into one garden, superior to whatever human hands could effect hitherto."

And for a few decades, a rapidly growing town in New England seemed as if it might prefigure that earthly paradise—or at the very least, a better industrialism.

chapter 3

The Utopia Commercial

WELCOME TO LOWELL, Massachusetts, 1833—glittering jewel of early American industrialism and home to its largest factories. Imagine it is quitting time—the fourteen-hour day over. Thousands of women spilled out of the factories, walking two at a time with interlinked arms. Their dresses and faces were clean. They wore bonnets and twirled green and blue parasols as they streamed out of red-brick textile mills as big as the White House.

Lowell was a "commercial Utopia," a vision of industrialization minus most of the bad stuff. The city of spindles was one of the most famous cities of the nineteenth century, and visitors came to see it from far and wide, including Abraham Lincoln, Charles Dickens, Henry David Thoreau, Nathaniel Hawthorne; even Southerners like David Crockett; and a wide array of foreigners. One historian wrote that "They rejoiced in the prospect of a clean, prosperous, virtuous factory life that should stand out splendidly against the grime and poverty of the great cities of England."[1]

In an America trying to make sense of what it was becoming, Lowell was a living, breathing model of what the country could be. In the new world, industrialization, nature, and society could live in peace. In just a few decades—overnight at the time—Lowell grew from a few farms into the second-largest city in Massachusetts.

Lowell, like the Erie Canal before it, was not just a massive engineering project. It was also part of an attempt to build a uniquely American "moral machine."[2] Much attention has been given to the employment model at the mills of Lowell. In an effort to avoid the ills of a permanent working class, young women from the countryside were supposed to cycle through the factories for a few years and then return, more worldly and more wealthy, back home. But waterpower was arguably as integral to the success of Lowell as being a happy place in the American imagination. The lack of smoke and soot was a key element of the place's appeal. "On approaching Lowell, I looked in vain for the usual indications of a manufacturing town with us, the tall chimneys and the thick volumes of black smoke belched forth by them," wrote one visitor. "Being supplied with an abundant water power, it consumes but little coal. . . . On arriving I was at once struck with the cleanly, airy, and comfortable aspect of the town."[3]

And one British visitor averred, "There is no steam-power there, and consequently little or no smoke is visible, and every thing wears the appearance of comfort and cleanliness."[4]

The utopian feel of the place coupled with the reality that people really still did have to work fourteen-hour days in clanging, loud factories got a young, abolitionist journalist, John Greenleaf Whittier, thinking about a man he'd once known. In Pennsylvania, he'd run into a "small, dusky-browed" German named Etzler who related to Whittier his "plans of hugest mechanism," whereby humanity would be restored to a kind of Eden.

> His whole mental atmosphere was thronged with spectral enginery; wheel within wheel; plans of hugest mechanism; Brobdignagian steam-engines; Niagaras of water-power; wind-mills with "sail-broad vans," like those of Satan in chaos. . . . By the proper application of which every valley was to be exalted and every hill laid low; old forests seized by their shaggy tops and uprooted; old morasses drained; the tropics made cool; the eternal ices melted around the poles; the ocean itself covered with artificial islands, blossoming gardens of the blessed, rocking gently on the bosom of the deep. Give "three hundred thousand dollars and ten years' time," and he would undertake to do the work.[5]

Greenleaf, just a few years later, could not help think: here was that hugest mechanism! A mechanism so large it portended an entirely new millenium, whether one liked it or not. "A stranger, in view of all this wonderful change, feels himself, as it were, thrust forward into a new century," he pondered. Peering over the town from a hill, he wondered what Etzler would make of the place.

> "Looking down, as I now do, upon these huge brick workshops, I have thought of poor Etzler, and wondered whether he would admit, were he with me, that his mechanical forces have here found their proper employment of millennium making," Whittier wrote. "Grinding on, each in his iron harness, invisible, yet shaking, by his regulated and repressed power, his huge prison-house from basement to capstone, is it true that the genii of mechanism are really at work here, raising us, by wheel and pulley, steam and waterpower, slowly up that inclined plane from whose top stretches the broad table-land of promise?

A hundred years into that millennium, Americans still haven't answered the question of whether—or perhaps which—technologies are raising us or lowering us. Are we closer to the promised land or farther away? How would we know it if we got there?

The brilliant fusion of the Scottish planned village, British industrial city, and the American utopian settlement[6]—the vision of a mechanized, profit-making near-utopia (where the girls were always young) was clearly a popular model for how America should industrialize. Perhaps the historian Caroline Farrar Ware overstated her case when she wrote back in 1931 that "the story of the New England cotton industry is the story of industrialization of America." Although there were other key events and places, other industries and forms of power, there was no single place that could be said to have had a greater impact in determining what industrialization *could be* to young America than Lowell.

Consider the alternatives. Small-scale manufacturing in the United States relied almost exclusively on water power through most of the nineteenth century.[7] The entrepreneurs and capitalists who would make themselves rich through mass production learned their skills by finding water power and exploiting it. What was new about Lowell was the unprecedented size and scale of its factories.[8]

The only other place where one might find such grand manufacturing was Pittsburgh, the iron city. Surrounded by rich coal veins, Pittsburgh is the yang to Lowell's yin. Because it was powered by coal, the city had to run its gas lights during the daytime, as the cloud of smoke and soot that hung over the city blotted out the sun. We don't have detailed measurements, but the atmospheric pollution around Pittsburgh during that time sounds as bad as any Shanghai horror story today. In fact, Peregrine Prolix, a wry, pseudonymous traveler from the South observed that "If a sheet of white paper lie upon your desk for half an hour you may write on it with your finger's end through the thin stratum of coal dust that has settled upon it during that interval."[9]

Pittsburgh was practically defined by its smokiness and sootiness. "Every body [sic] who has heard of Pittsburg, knows that it is the city of perpetual smoke, and looks as if it were built above the descent to 'the bottomless pit,'" a female traveler related.[10] The black clouds that hung over the city became a trope, a natural character to portray the city's Faustian bargain with its geology. In the Pittsburgh version of industrialization, dirtiness—not cleanliness—represented virtue, or at least money. "Its manufacturing powers and propensities have been so often described and lauded that we shall say nothing about them except that they fill the people's pockets with cash and their toiling town with noise and dust and smoke," Prolix pronounced.[11] Another traveler put it even better: "He whose hands are the most sooty, handles the most money, and it is reasonable to infer is the richer man."[12]

Conversely, Lowell factory girls went for walks in nature right outside the town, poking around the "rocky nooks along Pawtucket Falls, shaded with hemlocks and white birches." Along the way, "strange new wild flowers" were there for the taking.[13] Meanwhile, Pittsburgh was shrouded in a dark cloud of soot and ash. Its people were suffering, too. By the middle of the century, the American Academy of Arts and Sciences commissioned a report on ventilation and chimney tops, finding that the coal emissions had a deleterious effect on health. "The smoke, unless carried away and diluted in the upper regions of the atmosphere or consumed, is at all times injurious to health," wrote Morrill Wyman, a future physician to Teddy Roosevelt among others. "When bituminous coal is consumed . . . the black flakes of soot pass off from the chimneys and float in the atmosphere. This matter then falls upon the roofs of buildings and on everything else that is exposed to it. It enters houses, covers furniture, and soils clothing."[14]

Summing up Wyman's report's conclusion, a magazine editor wrote that "no one should be suffered to throw smoke or other impurities upon the atmosphere we 'breathe,'" thus backing Wyman's questioning of "whether any manufacturer or corporation should be allowed to produce such an amount of evil" as the air pollution was causing.[15] Still, at least one Pittsburgh doctor, who perhaps freelanced for cigarette companies in his off-hours, maintained that the soot and smoke "only go throat-deep" and, furthermore, that fire and smoke "correct atmospheric impurities."[16]

Compared with Pittsburgh or Manchester, in the early days Lowell was practically Shangri-La. "Nothing appears to be kept secret; every process is shown, and with great cheerfulness," Davy Crockett wrote of his 1834 experience. "I regret that more of our southern and western men do not go there, as it would help much to do away with their prejudices against these manufactories."[17]

Even Ben Brierley, a working-class Manchester writer who did not normally promote factory life, focused on the excellence of the environment for workers and other humans. "There's a bit o' comfort, becose we no' choked wi' soot, an' fluss, an' reech, an' bad smells, an' a general thickness o' air," he wrote through his dialect character, old Ab.[18] Thanks to waterpower, the air was nice, the houses were a clean white, the trees were healthy, and the sky was blue.

chapter 4

Prescribing for the Globe Itself

L OWELL'S CLEANLINESS did not come without a price, however. The owners of Lowell's factories, a loosely affiliated group of
Yankee businessmen known as the Boston Associates, had to permanently alter the rivers they used for power. Over the decades between the town's founding and the middle of the century, they built an
enormous machine, a system that eventually spread over 103 square
miles of waterway. They saw nothing wrong with their use of the river's
water as a commodity for producing power.[1] One historian summarized that

> realizing that nature could not be depended upon for continu-
> ous water power at maximum capacity which was needed for
> full realization of their productive ability—summer droughts
> and spring freshlets interfering with that goal—the Boston As-
> sociates built a series of canals, trenches, and dams at the very
> headwaters of the Merrimack.[2]

The looms and the town were the front end of a much greater system that was directly connected to an entire region's hydrology, climate, topography, and geology. James B. Francis, head engineer of the
Proprietors of Locks and Canals Corporation, built a machine that was
one hundred miles long and contained millions of acre-feet of water.
Its sole purpose was to supply power to a series of waterwheels at Lowell and the other sites run by the Boston Associates. Far upstream,
Lake Winnipesaukee, the largest lake in New Hampshire, acted as a

huge storage battery of potential power that could be tapped to smooth out power flow, particularly in the summer months. When Francis sent a message that he needed more water, several days passed before it arrived in Lowell.[3]

The enormous contraption had parts, like any other machine, but bigger. They were dams, canals, new turbines, and gates. Some of the most important parts of the machine weren't mechanical but instead conceptual. The Boston Associates figured out rules for selling a certain volume of water to make its disbursement less likely to generate conflicts among mill owners. They successfully battled adjacent land owners when their new dams flooded the land owners' property and were aided by a legal system that didn't know how to deal with problems like fisheries' collapse caused by alterations to the river.[4]

Furthermore, the knowledge paid dividends. The new turbines that Francis helped design and test roughly doubled the horsepower that could be extracted from the same amount of falling water, which allowed Lowell and other waterpowered mills to remain competitive with the increasingly steam-powered regions of the country.[5]

It might have been a wonder of the age, but some, like Henry David Thoreau, didn't find the damming of the river or the clattering of the machines to count much as progress. Thoreau's 1849 book about a canoe trip he took ten years before on the Merrimack and Concord rivers impugns the Lowell factories for destroying the fisheries: "Salmon, shad, and alewives were formerly abundant here, and taken in weirs by the Indians," but changes to the river, especially the factories and their waterworks, had destroyed the continuity of its living systems.[6]

Thus, Lowell's cleanliness did not come without ecological consequences. The entire Merrimack river system was transformed into a hydraulic machine for the purpose of making cloth. As a result, fish stocks suffered, particularly sturgeon, when spawning runs were obstructed. Unlike older dams, the new industrial ones were much more difficult for fish to navigate: The split between natural and human uses of the waterway tilted precipitously toward the latter.[7] In the 1860s and 1870s the Boston Associates even explicitly took on the responsibility of managing the naturalized system they had created, attempting to replenish the fish stocks of the Merrimack.[8] Thoreau would deride humans' attempts to help the fish along as "phil-*anthropy*," emphasizing the human-centered framework it implied.[9] If Thoreau felt the Lowell factories had destroyed nature, what could he possibly have thought of Etzler's similar, more grandiose plans?

As luck would have it, Ralph Waldo Emerson recommended that Thoreau review an obscure, almost ten-year-old book by an odd German who was bent on transforming the world with renewable energy. Thoreau titled his review of Etzler's book, "Paradise (To Be) Regained." In it, Thoreau immediately attacks Etzler's desire to make the whole country "a garden," sensing in the man no sense that the "unimproved" world should be respected. Thoreau then asked,

> And it becomes the moralist, too, to inquire what man might do to improve and beautify the system; what to make the stars shine more brightly, the sun more cheery and joyous, the moon more placid and content. Could he not heighten the tints of flowers and the melody of birds? Does he perform his duty to the inferior races? Should he not be a god to them?"[10]

To Thoreau, nature knows best. In comparison, humans are mere dissemblers and bumblers. How could broken people fix nature? In response, contemporary historians Robin Linstromberg and James Ballowe wrote, "The essence of Thoreau's critique of Etzler's Utopia was that Etzler put the cart before the horse. Etzler wished to harness nature to man's work before man had succeeded in harnessing himself."[11]

As such, we can see Thoreau laying one of the foundations for the ethic of the environmental movement: It's people's relationship to themselves and nature that must change in order to solve environmental problems. Technological fixes are destined to fail because the problem is humans and their social relations; improved means put to unimproved ends yield only mass unimprovement.

In one section, Etzler imagined a proto-food factory in which "one or two persons" cook for the masses with "nothing else to do but to superintend the cookery, and to watch the time of the victuals being done, and then to remove them, with the table and vessels, into the dining-hall, or to the respective private apartments, by a slight motion of the hand at some crank.[12]" Thoreau mocked him, calling out for humans to use a different crank, the crank that would change human behavior:

"But there is a certain divine energy in every man, but sparingly employed as yet, which may be called the crank within,—the crank after all,—the prime mover in all machinery,—quite indispensable to all work," he wrote. "Would that we might get our hands on its handle!"[13]

Thoreau disdained the ideas of progress prevalent in his day. He didn't think the world was becoming a better place. Nature was being destroyed because humans couldn't live simple lives but instead reached greedily after things and status. He thought local and small scale, seeing the wonder of creation in just about any natural system. Over time, Thoreau's philosophy has deeply informed many variants of environmentalism: Conserve, be self-reliant, think local.

Thoreau habitually observed the plants and animals near Concord with scientific rigor. He sought to "know the species of every twig and leaf" in the area. In particular, he began to study "when plants first blossomed and leafed." Year after year he ran to "different sides of town and into neighboring towns, often between twenty and thirty miles in a day." He "often visited a particular plant four or five miles distant, half a dozen times in a fortnight, that I might know exactly when it opened," he recorded in his journal. Meticulously, he input these observations into a chart from 1852–1858.[14]

His observations eventually became one of the best records available to phenologists, or contemporary scientists who study seasonal changes. A once sleepy field, in the context of studying climate change it's taken on new importance because plants are remarkably sensitive to small temperature change. In fact, in 2008 a team of scientists took Thoreau's record and went out to Concord to see how things had changed. Because 60 percent of the region is protected from development, it formed an excellent case study in how climate change alone could impact a set of flowers.

The results, published in the journals *Ecology* and the *Proceedings of the National Academy of Science* were stunning.[15] Although it's been only 150 years since his stint at Walden Pond, over that time the average temperature around Walden has risen four degrees. This change has altered when the seasons begin and end. The plants that can adjust their flowering cycles easily are surviving, but a bunch of others are dying. The highbush blueberries that Thoreau picked during September at Walden Pond are now flowering twenty-one days earlier than they did in Thoreau's day; they are doing fine. Others, however, are not. More than a quarter of the region's flowers are gone because of climate change, and another 36 percent are in imminent danger of dying out.

All that destruction has been caused by a slight build-up of carbon dioxide in the earth's atmosphere. There, the carbon dioxide traps more heat from the sun and the planet warms up. Coal plants in China, cars in Caracas, land use changes in Brazil, corn farms in Iowa—all of these things contribute to deranging the atmosphere. A molecule of CO_2 released counts equally, no matter if it happens in the middle of the Amazon or in downtown Detroit. The researchers who studied the changes around Concord concluded, "Given that climate-influenced loss of phylodiversity has been so great in Concord, despite 60 percent of the area being well protected or undeveloped since the time of Thoreau, a more global approach to conservation prioritization is necessary to minimize future species loss."[16]

Although Thoreau found it silly that Etzler would try to "prescribe for the globe itself," today there is no other way forward to stave off climate change and the attendant warping of every single local ecosystem, no matter how the locals might care for and coddle it. Here's what climate change means: The global environment has become an unintentional "garden," and humans have to manage it. When Thoreau, standing on the banks of the river on which Lowell and industrialism were built, thought about the relationship between other living things and men, he mockingly asked, "Should he not be a god to them?" As if to answer him, longtime environmentalist and mischief maker Stewart Brand's new manifesto is "We are as gods and HAVE to get good at it."[17]

In other words, environmentalism has been forced to go global. There is no retreating to the forest and protecting your patch from mechanized civilization. Environmentalism can't be a protest movement, a form of societal disobedience for the affluent. Individual nations, let alone people, can't solve the problem.

If climate scientists are right—and they are within reasonable bounds of uncertainty—the entire world has to start cutting its carbon dioxide emissions within a few years. If peak oil experts are right—and they seem to be within reasonable bounds of uncertainty—we'll need new clean sources of energy to prevent a worldwide shift to dirty fuels like tar sands and oil shale. We could just ask that the world limit its energy usage, but who would we beg? To what body could we appeal to prescribe for the world?

It's this suite of problems—fundamentally ecological but practically political—that led Doerr to tell his audience of technoptimists at the TED conference in 2007 that the climate change goal had to be to "make the right outcome the profitable outcome and therefore the likely outcome."

Environmentalists have to be able to compete, corporations and all. Moral reform may happen, but that hasn't stopped carbon emissions from rising. The prescription for the world is, in Google's corporate formulation, RE < C: renewable energy less expensive than coal. Right or wrong, cheap things win, so clean energy has to get cheaper.

It's an ugly goal. There's no poetry to it. It is not based in what Melville, following the German poet Goethe, called "the all feeling" of oneness with nature. It seems like no way at all to be an environmentalist.

But high-tech, low-carbon technologies seem to be the only way to preserve Thoreau's flowers, even if it means building and maintaining naturalized energy systems like the one that powered Lowell and destroyed the Merrimack's fish stocks.

In the future, environmentalists may be able to trace the roots of their ideas back as much to Etzler as to Thoreau. Already, Silicon Valley advocates have changed the national debate around energy. The highest levels of the Obama administration regularly echo not just the lines of reasoning that John Doerr employed but sometimes his actual applause lines from the TED talk. The head of the Department of Energy, Steven Chu, is one of the Bay Area crew. He went to the DOE after heading the greenest of the national laboratories, Lawrence Berkeley, which perches in the green hills above the site of the radical protests of the '60s. While announcing new green-tech grant awards, Chu told an audience at Google that

> the transistor made possible modern computers, the Internet, and Silicon Valley. The hybrid strains of wheat and the Green Revolution helped us feed a growing planet. Linking our computers together through the Internet unleashed an Information Age—in no small part because of the great ideas that have come out of Google. We are here today because this place reminds us

that, occasionally, radical innovation can alter the landscape of an entire industry. And we're here to announce a portfolio of bold new research projects, any one of which could do for energy what Google did for the Internet.[18]

Compare Charles A. Reich's diagnosis of the country's problems in his best-selling book *The Greening of America*. "What happened to the American people? In a word, powerlessness," he wrote. "We lost the ability to control our lives or our society because we had placed ourselves excessively under the domination of the market and technology."[19] Technology could not have been the answer either for Reich or Barry Commoner, another best-selling author who wrote *The Poverty of Power*. Scale, in itself, was a problem for E. F. Schumacher and followers of his *Small Is Beautiful* ideas.

However, Steven Chu and his boss, President Barack Obama, are nationalistic, radical, and more than happy to work with—not against—industry. Their vision of the energy future does not incorporate the countercultural version of environmentalism; instead, it is fundamentally pro-market, pro-technology, and pro-American. "This is the nation that has led the world for two centuries in the pursuit of discovery," Obama told an MIT crowd in a highly publicized energy speech in October 2009. "This is the nation that will lead the clean energy economy of tomorrow, so long as all of us remember what we have achieved in the past and we use that to inspire us to achieve even more in the future."[20]

The vision Chu and Obama present is nearly techno-utopian. But is the rush to build a political coalition to combat climate change and convince investors that betting on green will be profitable overwhelming the lessons of economy and respect for the natural world that Thoreau taught? Or will it take a naïve faith in technological progress to save the flowers of Walden Pond? If profit and salvation part ways, which road will John Doerr and his brethren take?

This book aims to provide a kind of memory and conscience for green-tech entrepreneurs and those who will live with their legacy. In tossing out the hoary truisms of mainline environmentalism, let's not forget the key critiques of global industrialism that environmentalism's point of view provided.

The world does have limits, a carrying capacity for human beings. Many of them are being reached at breakneck speed.[21] And picking a direction in our energy dilemma won't solve all the social, water, pollution, biodiversity, and land-depletion problems that come with a human population that's doubled since the year John F. Kennedy was shot.

Political scientist Langdon Winner argues that Americans created one constitution out of the American Revolution and a second out of the infrastructure of the Industrial Revolution. The paper constitution was argued over and debated from first principles: social welfare, justice, liberty, and so forth. "We the People of the United States, in Order to form a more perfect Union," the Preamble goes, "establish Justice, insure domestic Tranquility, provide for the common defence, promote the general Welfare, and secure the Blessings of Liberty to ourselves and our Posterity, do ordain and establish this Constitution for the United States of America."

Conversely, our industrial constitution—the one that shapes our material relationships to each other and the environment—has received far less attention and ethical investigation. It has been shaped by politicians, engineers, homebuilders, and investors who only occasionally have thought beyond the narrow profit-making possibilities of a technology. "In the technical realm, we repeatedly enter into a series of social contracts, the terms of which are revealed only after the signing," Winner wrote.[22]

The Obama administration and a host of green-tech proponents call the drive for low-carbon, renewable power nothing less than another industrial revolution. There will be turmoil and unrest and failure in our energy system as we try to rewrite the constitution we signed in concrete and steel more than a hundred years ago and that burst into the sky and out onto the land. This book documents our country's previous attempts to secure a more perfect power in hopes that future ones will succeed.

What Was II.

chapter 5

Steam-Powered America

L OWELL AND THE DREAM of problem-free power that it represented faded after the Civil War. In its place, hard-slugging capitalism accelerated. Companies, factories, cities, engines, and farms got bigger. They scaled up. The entire country began to function as a single market, united by railroads, the dominance of the Northeastern economic system, and the resources of the West. Size began to matter and the dimensions of mass production came into being. America's love affair with technological gigantism got hot and heavy.[1]

Steam power in a particular form—the Corliss engine—was sweeping the country. A factory could be built anywhere: No longer did a factory need fortuitous geography and hydrology; instead, all it needed was a Corliss engine, a steady supply of coal, and something to make. By the end of the nineteenth century the cheap land, labor, and easy access to the continent's natural resources turned Western cities like Chicago, Cincinnati, and Detroit into manufacturing powerhouses.[2]

These cities looked different from those built before railroads and steamboats, and steam engines had been used to eliminate the sense of human scale and distance that had been created in the horsedrawn era. "As trains moved faster, geography seemed to shrink," historian David Nye wrote of the time:

> The space between the new steam cities was annihilated, reduced to a passing panorama behind plate-glass windows. The passenger soon learned that it was impossible to focus on nearby objects as the foreground was reduced to a passing blur.

33

Railway travel refocused the eye on the distance, and travelers lost touch with the landscape's sounds, smells, and textures.[3]

Expectations of power and its possibilities changed, too. If factory owners wanted to use animal or water power to drive a factory, they were limited by the amount of horses they could reasonably keep or the hydrology of their region. Coal from Indiana and Illinois was cheap, so they could burn a lot of it, but steam engines couldn't transmit power very far. The towns had to be dense. A single steam engine's energetic output could supply energy to dozens of productive stations in a factory. There were no automobiles. Mass transportation amounted to horse-drawn buses on rails and then, later, trolleys of various types. Because of this, workers and their families had to live pretty close to where they worked; the exurbs would not have made sense.

As such, steam power allowed large Midwestern cities to come into being. Between 1830 and 1870 Chicago grew from fifty people on the shore of a lake to a bustling three hundred thousand–soul metropolis. By the last decade of the century a million people lived in the windy city. No Merrimack River could have provided the power for such massive growth. With the geographical constraints of renewable energy removed, cities could be built where the raw materials of mass production were, these being trees, cattle, water, and iron.[4]

Stationary steam power was a city thing. Although it allowed factories to get bigger, it didn't make sense out in the hinterlands of what was still a pretty sparsely populated country west of the Mississippi River and east of San Francisco. A few steam-powered plows started to appear in the country's great middle, but the centralized distribution networks required in order to get engines, steady fossil fuel supplies, and electricity to rural areas wouldn't really develop until the rise of long-distance electrical transmission, oil, and automobiles in the twentieth century.[5]

chapter 6

The Wind and the West

WHEREAS COAL BOOMED in the cities, in the arid West, energy wasn't really the limiting factor. What mattered—and still matters—in what was once known as the Great American Desert is water. As steam power scaled up the Middle Western cities, windmills, built and maintained by entrepreneurs and settlers, removed a key barrier to settling the country's interior.

With a windmill, a family could move to a place where less than twenty inches of rain fell per year so the land was cheap. With some luck and hard work improving the land, perhaps they could sell that farm and buy a nicer one, moving up a notch in the social hierarchy and securing a more permanent existence.

That was the promise of the West, "the frontier!"—as men and women who had never been there heard it whispered in their ears. Some people even told them that the "rain follows the plow," and as more people moved west, the nice, rainy conditions of the Mississippi valley would go with them.

When their wagons and bodies met the actual arid land west of the hundredth meridian in western Kansas or Nebraska, most of them became more concerned with survival than prosperity. And as a survival tool, almost nothing could beat the light, cheap American adaptation of the old European wind machine. Each windmill not only represented a meager few tenths of a horsepower but also a life insurance policy. "It was the acre or two of ground irrigated by the windmill that enabled the homesteader to hold on when all others had to leave," wrote Walter Prescott Webb in his grand march across *The Great Plains*. "It made the

35

difference between starvation and livelihood."[1] Windmills also enabled the cattle industry to expand significantly. "Until their introduction, a rancher's grazing was limited by surface water: a cow will not amble more than fifteen miles a day for water," wind historian Robert Righter noted. "Thousands of square miles were unusuable for livestock-industry purposes." However, windmills pumping water into scattered stockponds opened up all that rangeland.[2]

Aside from the stove, the windmill was the most popular personal mechanical power source of the nineteenth century. Without it, the West beyond Wichita and Omaha could not have sustained the people that knit the nation together, however thinly, from sea to shining sea. It's impossible to know exactly how many windmills flowed into or were built in states like Nebraska and Kansas, but their numbers astounded visitors to the region. One early twentieth-century traveler counted 125 windmills in the small town of Colby, Kansas, alone.[3] "The prairie land is fairly alive with them," the *Kansas City Star* reported. "The windmill has taken the place of the old town pump, and no western town is complete in its public comforts without a mill supplying water to man and beast by energy of the wind." As windmills came into greater use, Nebraska's population jumped from 123,000 to more than 450,000 in the 1870s. Over the same period the Kansan population jumped from 365,000 to almost a million. "[The windmill] was not only a convenience, but a necessity," Webb wrote. "Without it large areas would long have remained without habitation."[4]

The windmill's stunning success, however, was not a result of technical efficiency. Many of the windmills used during the time period did not convert much of the wind's energy into power. In fact, expensive models were sometimes worse than those built in the hinterlands by a farmer and his children.[5]

The eminent historian Nathan Rosenberg argued that it wasn't strictly horsepower-for-horsepower competition that determined the usability of a power technology but rather a much broader set of factors.[6] Windmills could not have competed with steam engines or water turbines in driving a factory, but that wasn't their competition. Instead, the windmills were competing against aridity itself—and by drawing on the resources of the underground, the windmills often won. The humans on the plains voted with their labor and their money: More than six million water-pumping windmills operated in America, one scholar estimates.[7]

The size of the market for them at the beginning of the 1880s brought the incredible optimizing forces of money and science on this homegrown technology with tremendous results. Before the 1890s intuition and luck were as important as math and know-how. The widespread use of the windmill outstripped the scientific understanding of how best to capture its energy for human purposes.

Consequently, someone had to figure out how to take all the folk wisdom about windmills and transform it into a better product. Someone had to figure out how to control the conditions in which the mills worked, to hold the wind steady so that the machines could be measured. It wasn't easy.

TAKING THE WINDINESS OUT OF THE WIND

The brick building was thirty-six feet wide and forty-eight feet long, with high ceilings that stretched nineteen feet to the roof. A fourteen-foot wooden beam hung from the ceiling attached by pulleys, belts, and gears to an eighty-horsepower steam engine. When the machine was turned on, the beam rotated, cutting a perfect circle exactly one-sixtieth of a mile at the outer edge. That clever choice meant that the number of turns the arm made per minute was equal to the speed of the arm in miles per hour.[8]

This strange contraption was a secret artificial wind machine designed by Thomas Perry in the spring of 1882 in the factory yard of the U.S. Wind Engine and Pump Company, a leading windmill manufacturer. The company was the product of Daniel Halladay's invention of a self-governing windmill decades before combined with the business acumen of Daniel Burnham along with the money of the wealthy residents of Batavia, Illinois, the Chicago suburb that later became home to the world's most powerful atom smasher.[9]

Perry's secret tests tended toward the elegant. Four iron straps allowed Perry and his helpers to attach any of fifty different windmills, each five feet across, to the end of the apparatus. By running each wheel indoors, under the most controlled conditions possible, he'd done the nearly impossible: He'd taken the unpredictability of the wind out of the equation. In Perry's model, the "artificial wind" always blew at 8.5 miles per hour, the work the wheel had to do was constant, and the windmill was always five feet across. The only variable, then, was the design of the wheel itself and how its blades caught the wind.[10]

By applying a given amount of braking force to the wheel and then counting how many times the wheel turned per minute against that load, they could measure the force of the windmill. Perry was a careful man. Each wheel design he tested was given multiple test runs to attain its average performance and he ran a control windmill dubbed "#2" each time to make sure that weather conditions or unknown variables were not negatively impacting his testing.[11]

Over the next year he put the dozens of windmills through five thousand runs, calculating averages, making tables, and systematically pushing his way to a new understanding of how best to catch the wind. A miniature industrial research laboratory, like Edison's Menlo Park or Bell Labs, was born in Batavia—and even without basic aerodynamics knowledge, the tests led to the radically more efficient windmill design. Even in today's wind industry, the experiments are considered remarkable.[12]

The doors to the testing room were kept closed and guarded to prevent crosscurrents. But Perry's knowledge had also become a valuable thing, so perhaps his superiors realized they should keep prying eyes away from such sensitive experiments. After all, there was fierce competition in Batavia between Perry's employer and the Challenge Windmill Company. The tension in the city reflected a wider shift in the windmill business. It'd gone national. In the wake of the Civil War, more than one hundred manufacturers got into the windmill game.[13]

Prior to the 1870s only a few companies had dominated the trade. Halladay is generally credited with building the first windmill that could furl its own "sails" back in 1854. Its wooden blades were tacked to a rim on the edge of the wheel. Under standard winds, its force simply turned them as it hit the blades of the machine. But if the wind blew too hard, the blades moved on their hinges almost like a Mai Tai umbrella, turning the flat-face of the wheel into a cylinder. Because the blades were attached to a rim around the outside, not in the center, the windmill ended up being shaped like an open cylinder, which allowed the wind to blow right through it.[14] With this innovation, the career of the relatively cheap and light "American windmill" had begun.

The self-governing windmill seemed like a technology that could find a market, but it wouldn't be in the East. John Burnham, who had encouraged Halladay's pursuit, decided that the real market was in what was then called the West: Illinois, Ohio, Missouri—anywhere west of the seaboard. So Halladay, Burnham, and what became the U.S. Wind Engine and Pump Company moved to Batavia to set up shop in 1863. Local town boosters gave them some money and a good deal on a brick factory near town. The company had serious skin in the game. It was possible they had also become the dominant supplier to the railroads, who needed water for their steam locomotives; the ranchers, who needed the water for their cattle; and the settlers, who wanted to live.

But a competitor emerged from the backwoods of Wisconsin. Reverend Leonard Wheeler, a missionary among the Ojibwa nation, invented a new type of windmill in 1867. It looked like the Halladay, but it had a new—and ultimately much more successful—way of furling and unfurling its sails. Instead of Halladay's creaky joints, Wheeler's windmill had a vane jutting off to the side. The wind struck the vane at a perpendicular angle to the windmill, which tended to push the face of the wheel away from the wind. To compensate, a weight was attached to the vane so that it wouldn't throw the wind machine out of the wind until the force of the air was going to bust the mill. From the wind's perspective, a Wheeler mill facing it was a great target, but when thrown out of the wind, it was barely there.[15]

The system was a bit janky, but in an ingenious way. Nearly all future factory-made mills adopted some variation of the side vane, though a few vaneless models were still produced. The Eclipse became the dominant mill used by the railroads, who often needed massive wheels two car-lengths wide. Halladay's company was forced to put out a version of the solid-wheel windmill in later years.

By the end of the 1870s every manufacturer trumpeted some advance or another that distinguished it from the competition, but no one really knew if they were telling the truth. All kinds of windmill blades were in use. Sometimes, there were sixty thin sails clustered together; other times, just a few big ones. They were set at all kinds of different angles. Some stuck up at nearly 40 degrees from flat, others at 20 or 30 degrees. There wasn't much rhyme or reason to these decisions. Inventors were guided largely by intuition or habit. If it worked, it worked. If it broke, they fixed it. Everyone tried to make the best mill they could, but they continued to rely on their sense of technological aesthetics or on the folk wisdom of how a windmill should work.

Wind is not an easy power for which to develop rules of thumb. Although its power and direction could be generally known, determining what it might be up to at any given second seemed as poetically hopeless as measuring a flickering flame. The wind's streakiness, as an old teacher of Thomas Perry called its tendency to come and go, made it a troublesome input in any experiment. The wind is also three dimensional; it doesn't blow in straight lines, as companies continue to find out even today.

Before Perry, an English civil engineer, John Smeaton, had conducted the most authoritative tests on the properties of the wind as it related to generating mechanical power, and he delivered his findings in two lectures before the British Royal Society in 1759.[16] A few scattered attempts followed, but as Perry noted, Smeaton's work had "remained the only definite available source of information on the subject treated" for more than 120 years! (Perry himself had made sure to track down a copy of the work before beginning his experiments.)

For people who could afford windmills—mostly the cityfolk and wealthier farmers of the country—the profusion of windmill companies, wide variability in their designs, and lack of a consistent way of rating them put consumers in a tough spot. Windmill companies all claimed some breakthrough, yet there was no government or industrial body that could certify those claims until the late 1890s. A company could attach a number to demonstrate how well a windmill worked—how many gallons of water it pumped—but then there were all sorts of other variables that defied traditional scientific testing. There were just too many variables that went into the success of a windmill; it was impossible to establish causality. The wind varied from place to place. There was the streakiness of the wind, how often it gusted, whether it blew true or was turbulent, and whether it approached at funny angles. More variables could play a role, too, like the barometric pressure during a test. Then there was the height of the tower or the efficiency of the pump and how far down it was drilled and into what type of aquifer. The list went on and on.

Even measuring how hard the wind was blowing with the anemometers of the day was tricky. All kinds of adjustments had to be made for them, particularly in the stronger winds, when windmills could do the best work. Under the conditions, it was nearly impossible to figure out exactly whose windmills had the right stuff and whose were junk.[17]

There was one key variable, though, that was under human control: the design of the sails of the mill. Perry figured out that if he curved the sails a bit and set them at just the right angle to the wheel, the mill could capture far more of the wind. Build it out of steel and it could spin much faster. Slap some gearing on the windmill to turn its quick movements into slow, powerful strokes of a pump and it was an entirely different class of machine. It was both better and could be manufactured at far lower cost than the mills of the day.[18]

After the tests concluded in 1883, his brain brimming with his new ideas, Perry presented his results to the board of the U.S. Wind Engine and Pump Company. One can imagine how excited he must have been to present what he knew to be a revolution in windmill design. The management, though, wasn't swayed by his fancy scientific methods. "Its board of directors rather curiously declined to accept his recommendations for redesigning its products," wrote the eminent windmill historian T. Lindsay Baker. "The very conservative U.S. firm thus refused to accept the results produced by science."[19]

Luckily, there was other financing available for his ideas. LaVerne W. Noyes, who had developed new agricultural tools and—more famously—a best-selling dictionary holder, bought into the concept. Together, they founded the Aermotor Company in 1888, which eventually became the most successful windmill maker in the world.

That the company would come to dominate half the windmill market by 1900 was not immediately clear. The first year, they managed to sell only a measly forty-five windmills; nonetheless, they grew quickly. Detailed records of the company's sales aren't available, but by 1891 they were selling tens of thousands of windmills annually. Even in 1950 the company claimed to have eight hundred thousand windmills in service, with "more than half" in service for more than forty years. They crushed competitors, "pricing them out of the market through its incredible efficiency of production," thanks to Noyes and their wind-mills' excellent performance.[20]

By any measure, therefore, the Aermotor was an incredible success that deserves a place among the major innovations of the late nine-teenth century. Perry married scientific rigor with a half-century's worth of inventors' ideas.

PERRY'S EMPLOYER, the U.S. Wind Engine and Pump Company muz-zled the publication of his data for seventeen years, long after he'd gone on to cofound the Aermotor company. It wasn't until 1899 that his re-sults were publicly available. In the intervening years scores of settlers had poured into the arid West, sparked by the various Panics (capital P) that struck the Eastern economy. Windmills followed them like the dust and ashes of dashed frontier dreams.

THE DIY WINDMILLS OF THE ARID WEST

Between 1884 and 1887 "a torrent of immigrants settled the western third of Nebraska and Kansas." During that time, Nebraska and Kansas happened to be unusually rainy.[21] As humans had recently pushed into the area, a gaggle of human-centered theories sprang up about what could be causing the "trend." Some people suggested that the "iron on the railroad lines" or the "wires of the telegraph lines" were responsible. Others thought "the disturbance of the atmospheric circulation through the concussions of locomotives and moving trains" caused the phenomenon. More widespread was the idea—born in conservationist circles—that "forests produce rains."[22]

But an eminent Nebraska natural scientist, Samuel Aughey, looked at the tree planting data and noted that the rains began before the trees. His natural conclusion was that it must be settlement. "There is, how-ever, another cause, not heretofore mentioned, most potently acting to produce all the changes in rainfall that the facts indicate have taken place. What then is that cause?" Aughey wrote. "It is the great increase in the absorptive power of the soil, wrought by cultivation, that has caused, and continues to cause an increasing rainfall in the State."[23]

Now that had a nice ring to it. By farming the land, settlers made the climate around the land better. That's a theory on which people can hang all kinds of real estate speculation—and so they did. Charles Dana Wilber, a promoter for the Burlington Railroad and the great state of Nebraska, gave the arid West's boosters a catch phrase summarizing Aughey's work: "Rain follows the plow," Wilber declared.[24] He was quite amazingly wrong. As the dust bowl of the 1930s would show, drought followed the plow, as small farmers proved unable to keep erosion in check themselves or band together to reduce those problems as they did in later years.[25]

Regardless, railroad boosters took that catchphrase into the cities of Reconstruction-era America to tout the Great Plains. In the process, of course, their own land holdings and business operations would get much more profitable. Albert N. Williams, in *The Water and the Power*, stated,

> The rail companies, attempting to build up the regions which they served, put a battalion of high-pressure salesmen into crowded and troubled Eastern cities, there to shout the wonders of the Great Plains: Kansas and Nebraska, for the most part. These schemers carried with them artists' conceptions of farming in that dry and dusty region—wonderfully colored prints of farmers standing beside twenty-foot corn stalks, hundred-pound squishe, or whatever the plural of squash is, melons which would break a man's back in the picking, and other agricultural wonders.[26]

The boosters had a problem, though. People had heard that the West was as dry as a bone. John Wesley Powell had told them in 1869 that anywhere west of the hundredth meridian and east of the Pacific Coast was just too damn dry for anyone to seriously consider living and farming.[27] And that's where Wilber's work came in handy: Why, sure, it *had been* too dry . . . but settlement was fixing all that. Williams continued,

> When the normally cautious New Englander would suggest that he had heard from travelers that the Plains were as dry as a bishop's cupboard, he would be told either one of two fantastic fictions: that the very act of plowing up the unbroken sod made it rain; or that there was a strange natural phenomenon taking place, a "Rain Belt" which was moving eastward from the Rocky

Mountains at the rate of twenty miles a year, and that the canny investor would do well to put himself in hock for as many parched acres as he could swing, after which he had only to wait a short time until the Belt would arrive, upon which occasion he could subdivide and sell to less foresighted individuals at a fat and un-Christian profit.[28]

It was a good sales pitch and it worked like a charm. People came from all over. Then, the rain stopped and the region regressed to its climactic mean. About half the settlers of western Kansas tucked tail and ran back East from 1888 to 1892, painting their wagons with bumper sticker–like slogans such as "In God We Trusted, In Kansas We Busted."[29] Some turned to "rainmakers" who fired explosions into the skies or cooked up special chemicals, usually including sulfur from traveling wagons.[30]

The practical among them who still had money left bought windmills to bring up the subsurface water. The rest—broke, starving, and nearly out of water—built their own. Webb wrote,

> Some held their ground either because they were unable to move or because they had found a way to live with the country. These windmills of Nebraska were not those factory-made rosettes which you still see whirling so lightly on your western horizon. Few of those people had money enough to buy a windmill. Just as the people had to learn to make fences without rails, so they had to learn to make windmills without steel and without money. And there sprang up from Lincoln to the western edge of Nebraska and on into Colorado on both sides of the Platte Valley the greatest aggregation of homemade windmills known in all history.[31]

Homemade mills could look like just about anything, from a kind of overshot setup that looked like an old fashioned waterwheel, to the Battle Axe, which had two axe-like blades that chopped the air, to the more conventional type with a flattish fanlike face and some number of blades. Every town had its own style, usually modeled after the style used by first guy who put one up.[32] That neighbor was likely to know the local conditions, materials, and water depth. The proof of the windmill was in the pumping. A person could go to his place and take a look. If his mill was working, then you knew it worked. If it didn't cost much, raised water, and didn't break down, it was doing its job. The point wasn't to get the most horsepower or the most torque or the most efficiency; instead, the metrics for success were much more human and fuzzy. There was no comparison between models, no tables of competing performance criteria. What mattered was that wind machines were "the cheapest power the farmer can employ on a farm," as an 1892 farming manual put it.[33]

They could be built out of anything for nearly nothing. Erwin Barbour, the Nebraska state geologist who took a special interest in homemade windmills, wrote,

> Old wire, bolts, nails, screws, and other odds and ends of hardware, old lumber, poles, and braces such as are common to every farm, enter largely into construction. Even neglected mowers, reapers, and planters, old buggies, and wagons contribute material. . . . The farmer who is inventive enough to build a mill is competent to see quickly the adaptability of certain parts to his ideas. It is this use of old and neglected material which is particularly recommended in this connection, for in making a mill of low efficiency, such as most homemade mills are, cheapness is the main object. Many mills have cost nothing whatever. Others cost $1, $2, and $3 and occasionally as much as $50, $75, and even $150. . . . The writer considers $3 a liberal allowance for everything needed on an ordinary farm for the construction of a strong, satisfactory, and lasting mill.[34]

Barbour, an East Coast gentleman educated at Yale who ventured West into the plains beyond the hundredth meridian, found the windmills an awesome sight. Amidst the existential flatness and desolation of the prairies, he found proof of man's glorious ingenuity, even the man of a decidedly baser instinct. In town after town, the settlers, unable to afford factory-made windmills to transform wind into water, had built hundreds of homemade contraptions doing the same. The average farmer, far from being a sadsack dullard barely making ends meet, was as inventive as a Yankee—and under much tougher conditions.

Barbour is not the type of man one would expect to find promoting cobbled-together machinery. The Nebraska state geologist, he expected a certain level of decorum from himself and others. He dressed impeccably in three-piece suits and wore two pince-nez attached to his person by a gold chain. A paleontologist by avocation, Barbour was, first and foremost, an eminent fossil hunter. When working on specimens, his underlings could let loose and remove their suit jackets—provided they retained their neckties.[35]

In composing government documents, he never strayed from the formal third person. He composed two such resources for the U.S. Geological Survey on the DIY windmills of Nebraska. "Those who have had little chance to observe for themselves can scarcely be brought to realize the great number of homemade mills, and the wide territory they cover," Barbour wrote in a U.S. Geological Survey report. "But to the writer's knowledge they extend in almost unbroken succession from Omaha to Denver, and from South Dakota through Nebraska, Kansas, and Oklahoma."

> Barbour would know. He'd sent a team of three graduate students gallivanting all over the state of Nebraska documenting the profusion of windmill styles. He popularized the fun local names for wind machine variations: Jumbos, Merry-Go-Rounds, Battle-Axes, Mock-Turbines, Dutch.

Barbour continued,

> The writer has gone by rail over the various roads; has driven a team; has employed—at his own expense—students to drive several times across the State in various directions. In this way as will be very plainly seen, a large number of places have been visited, and a very fair survey of the windmill has been made, and from the knowledge obtained it may be said that the Platte valley from Omaha to Denver seems to be the very backbone of the homemade mill.[36]

Though windmills blanketed the countryside, they were built out of desperation and on guesswork and hunches. Although they worked, they didn't work as well as they could have. For instance, a huge jumbo—the machines shaped like a big water wheel—delivered about four times less power than a comparable Aermotor.[37] "Homemade mills are, of course, of low efficiency from a physical and mechanical standpoint; yet they are capable of doing all that is demanded and more," Barbour wrote in the delightfully strange United States Geological Survey pamphlet, "Wells and Windmills in Nebraska," which investigated the DIY mills of his adopted state. "They cost little, wear well, and do all the work that is laid on them, so that it makes little practical difference whether some of them are of low or high efficiency."

However, local improvements could not spread. Knowledge was passed by word of mouth, and the communication networks that could have amplified good technical signals and reduced noise in windmills were nearly nonexistent. Every town had backers with investments in their particular civic unit and the set of people and technologies that sat near it. There was a lot of experimentation, but little precious learning.[38]

In some ways the problems they encountered foretold what would happen to the "appropriate technology" movement of the 1970s. It took—and still takes—larger agglomerations of time, capital, and expertise to make wholesale improvements in technologies. Simply working out and manufacturing good mills in small communities with few resources is difficult.

chapter 7

The Parable of Petrolia

WHEN PLAINS SETTLERS needed every kilowatt-hour they could get, a new and savage energy source—crude oil—was flowing out of new wells in the nation's first oilfield in western Pennsylvania. The only problem was that sometimes the petroleum didn't flow fast enough. Regardless, early oil services companies run by Civil War veterans had a solution.

On January 21, 1865, Colonel E. A. L. Roberts carefully loaded eight pounds of black powder into an oblong iron casing, affixed a cap as a detonator, and lowered the charge on a wire down into the Ladies Well near Titusville, Pennsylvania. After getting to the depth he thought proper, Roberts took out a donut-shaped weight and sent it down the wire. According to plan, the "torpedo" blew up, letting off a terrible report and sending debris flying high into the air. Then the Colonel, who gained his explosive expertise in the finally waning Civil War, did it again.

Soon thereafter, what had been a depleted well started coughing up "a steady stream" of black gold. Roberts was in business. The next month he set up a business charging up to $200 per well and a royalty of one-fifteenth of the increased flow of the oil. "Shooting the well" became a standard practice at the world's first major oil field. The rule was capture as capture could, and Roberts's torpedoes opened up your whole underground regions for extraction — whether or not they were located under your particular property.[1] The Oil Creek Valley, about one hundred miles north of Pittsburgh, was the center of the oil business from 1859 to 1873.

Oil was an alternative energy source then—a strange, mysterious substance in a world still using wind and water, wood and coal. Oil didn't send cars zooming around or get turned into plastics back then. People all over the world burned it in lamps as a replacement for a set of illumination alternatives that weren't quite right for the task. There was whale oil, but that was getting tougher to find. Whalers spent more and more months farther and farther away from population centers to fill up their barrels with bounty from the most majestic creatures in the world. And even if they killed a hundred whales, they were bringing back only a few hundred barrels of oil. There was pig-derived oil, too, along with gas made from coal. Historian Brian Black has found that there were already fifty-six coal-to-gas plants operating by 1850.[2]

Furthermore, new lamps introduced in the 1850s allowed consumers to burn pretty much anything they wanted, decreasing the cost of switching fuels. Black went on to note that after that

> each illuminant helped bring light to darkness. However, each product left dramatic room for improvement. While each development functioned to lay groundwork for the rapid acceptance of petroleum upon its "discovery," the coal oil industry, which grew significantly in the United States during the 1850s, achieved a national distribution network that could be shifted most easily to other fluid energy commodities.[3]

So there was a market and ecosystem awaiting the product that could fulfill "the divine potential of increasing time in the day." Some people had discovered that "rock oil" could be distilled, just like whale oil, but it was too difficult to collect where it seeped up to the surface. People sometimes skimmed the crude from the surface of the waters where it naturally got stuck or they sopped it up with blankets. Some used it as a tonic: They "drank freely of the water, which, by and by, 'operated as a gentle purge.'"[4] Not exactly the way to go from rags to riches.

Nonetheless, some Yankee capitalists from Connecticut were convinced that oil could be found in the ground and exploited. They recruited "Colonel" Edwin Drake, who was not a colonel at all, but got away with the name mostly because he was charming and unknown in the region. He, in turn, found someone skilled in the art of drilling, or what passed for it in those days.

Drake and his sidekick "Uncle Billy" Smith started looking underground for oil in the spring of '59. They used a heavy metal tip attached to a rope, sending it plummeting down the borehole like a ram to break up the rock. It was slow going. Working with a local machinist, they simply pounded a hole in the ground with a heavy piece of metal attached to a rope threaded through a pulley to a steam engine. It took weeks of "chipping" to go the sixty-nine feet down to the reservoir. On August 27, 1859, at sixty-nine feet of depth, Drake and Smith hit oil.[5]

"We have no language at our command by which to convey to the minds of our readers any adequate idea of the agitated state at the time we saw [the well]. The gas from below was forcing up immense quantities of oil in a fearful manner and attended with noise that was terrifying," wrote Jim Burchfield, editor of the *Titusville Gazette*, on seeing one of the first wells ever sunk. He went on,

> When the gas subsided for a few seconds, the oil rushed back down the pipe with a hollow, gurgling sound, so much resembling the struggle and suffocating breathings of a dying man, as to make one feel as though the earth were a huge giant seized with the pains of death and in its spasmodic efforts to retain a hold on life was throwing all nature into convulsions.[6]

Though oil had been drilled for in Canada before, Drake's well and the huge industry it portended really did change everything. Oil wells pumped an immediately salable commodity out of the ground. Suddenly, a piece of land wasn't just its acreage, but its volume. What was under one's house was quite likely to be more valuable than the house itself. If windmills turned air into water, oil wells turned earth into cash. A photo taken a few years later shows a simple well. Its caption reads, "Source of the world's most gigantic fortunes—pumping wells in the oil country—western Pennsylvania."[7] Indeed, just in 1870, before the area was producing at full capacity, oil wells made their owners and operators $19.3 million.[8] Not bad for a backwater part of the country that even now remains largely undeveloped.

The end of the Civil War quickened the pace of development. Veterans returning from the front went out looking for places to make money, and they found them in places like Pit Hole, one of the fields discovered in 1865. During its time at the top, the Oil Creek Valley, which became known as Petrolia, gave up fifty-six million barrels of oil. As more people poured into the oil regions and more and more torpedoes got sent down wells, "supply outran demand" and soon the whiskey barrels that held the oil "cost almost twice as much as the oil inside them."[9]

The cheap supply let inventors experiment with a tremendous energy source. Beyond the money the good wells were making their owners, they were also extracting an incredible amount of energy from the ground. One team of researchers have estimated that for every unit of energy early oil prospectors invested sinking a well, they got back "more than 100 times as much usable energy."[10]

Oil was—and is—uniquely convenient. It is what's called energy dense: To equal the amount of energy in a tank of gasoline, you would need to burn two hundred pounds of wood. It's stable, meaning it doesn't explode randomly. And it's a liquid, which means that you can store it in barrels or tanks and transport it in a pipeline. It was the killer app for the infrastructure age.

Thus, the American fossil-fuel era had begun. And then, just as quickly as it burst onto the scene, the oil fields in Petrolia started to give out. The brutal logic of the rule of capture, a glorified "finder's keepers" for oil, meant that everyone was incentivized to pump and torpedo as much as possible in order to just get the oil first. It's a blindingly simple rule that drove the entire development of the region. If you could pump it, you could sell it. So people pumped as fast and as hard as they could. And then the forests of derricks pumping slowed to a halt. People left. Towns died. The trees rushed back in on the place.[11]

Petrolia, the river valley in Pennsylvania, was done for, but Petrolia, the American condition, was beginning. Huge oil field finds in Texas and California in the early twentieth century meant that America could basically treat oil as free for the next seventy years. By the time oil surpassed coal as the most burned fuel in the country, Americans had become dependent on the strange, energy-dense liquid created under unusual conditions millions of years ago.

The change wasn't immediate. The best days still lay ahead for windmills and wave motors and horses and coal. But Petrolia lit a fuse that would run through the entire century until it exploded in 1973 during the OPEC oil crisis. During that time Americans realized they had built a country on a fuel source that was no longer coming up from the domestic geology. Finding oil had become increasingly more difficult. And the world is now experiencing the same thing. Oil usage has outpaced new finds for decades. No one can say exactly when, but some day soon the world's oil production will begin a long decline.

chapter 8

Wave Motors and Airplanes

A YOUNG MAN WITH ARTISTIC aspirations could not have resisted the crowds of Market Street on a Saturday night. Nothing was more San Francisco than the street that cut through its heart. Like a weekly fair, all classes of society and the many flags of a port town mixed on the promenade from Powell to Kearny. "Everybody, anybody, left home and shop, hotel, restaurant, and beer garden to empty into Market Street in a river of color," wrote one young woman of the time.[1]

Among the throngs of sailors and servants, we could almost certainly have found a young Jewish kid with an overbearing father and a canted, humane take on human foibles. Long after the 1890s and far away from the city by the bay, he would make a name for himself with a set of drawings that made him the most popular cartoonist of the machine age.[2]

It's certainly not much of a stretch to imagine the twelve-year-old Reuben Goldberg participating in the weekly Saturday night parade and happening past a working model of one of the oddest machines he was likely to have encountered on the foggy streets of the city. The Wave-Power Air-Compressing Company was one of a half-dozen concerns that were attempting to harness the waves of the Pacific. And it just so happened to have an office at 602 Market, just a block from the main San Francisco procession. It may have been the sort of place that a machine-obsessed little boy might have found himself wandering on a Saturday night.[3]

There he might have seen the small model that the company invited the public to come inspect. To the untrained eye, it might have looked like a very complex pier. A float attached to the structure could move up and down freely as the operator raised or lowered the level of water. Atop the pierlike contraption, there would have been a series of pipes containing compressors hooked onto a reservoir for the pressurized air.[4] The machine's inventor guaranteed that "whatever the extent of the perpendicular movement, the pumps take in some air and effect some compression, and thus do some work." From there, the promoters of the company would have told anyone who cared to listen that the compressed air could be piped to shore, where it could run dynamos to generate electricity.[5]

Like the other wave motors of the time, the model machine purported to show, step-by-step, how the horizontal or vertical motion of the waves would be converted into usable power for human beings. And always, this seemingly simple transformation seemed to require an inordinate amount of pumps, and chambers, and floats, and levers, and pulleys.[6] They seem like terribly serious versions of what has come to be known as Rube Goldberg machines. The adjective derives from an insanely popular series of drawings Goldberg did in the 1920s called "Inventions." One can now use his name to describe "any very complicated invention, machine, scheme, etc. laboriously contrived to perform a seemingly simple operation."[7]

One exemplary Goldberg cartoon shows how to build a better mousetrap, the constant aim of American inventors. In it, a mouse dives for a painting of cheese but instead breaks through the canvas, which lands him on a hot stove, so he jumps off it onto a conveniently located block of ice that is on a mechanical conveyor that drops the mouse onto a spring-loaded boxing glove that sends the mouse caroming into a basket that triggers a rocket that sends the mouse in the basket to the moon.[8]

There's a curious resonance between Goldberg's famous cartoons and the wave motors of the 1890s. In both, there are no black boxes. Every part, in one way or another, has to physically touch every other part. Electronics didn't exist and dynamos would ruin the fun. But if the classic drawings gently mock the foibles of mad inventors, it's in the wave motor inventors of *fin-de-siècle* San Francisco that Goldberg could have seen the dead-serious version of ill-fated mechanical creative obsession.

The group behind the machine might have been delightfully zany to the young Goldberg, too. The company was the brainchild of Terrence Duffy, an inventor who had recently completed a self-published book called *From Darkness to Light: Or Duffy's Compendiums of Nature's Law, Forces, and Mind Combined in One* (1893), which purported to explain all the mysteries of nature through magnetism. It served up wisdom like, "The blood is a magnetic fluid, floating in the tension of the body. The brain is the equivalent to a magnetic or electrical storage battery or coils. The brain floats in the tension of space, each organ being like millions of fine wires coiled in receptacles, for the storage of impressions, or experience, or intelligence."[9] A later book received a rather discourteous reception in the *San Francisco Chronicle*, in which the reviewer wrote, "mental unsoundness is everywhere visible in this book."[10] However, the only non-wave-motor or book-related mention of Duffy in the San Francisco papers of the era was his wife's 1888 (very) public appeal that he properly support his three children.[11]

But even if he was a deadbeat dad and a bit of a nut, Duffy had a dream as big as the Pacific Ocean and little could deter him. As a result, the Wave-Power Air-Compressing Company was incorporated in May of 1895. A florist-cum-inventor, Duffy, along with a small group of friends, offered a million dollars of stock. That is to say, they created a million shares out of thin air and offered them at $0.25, far below the "par value" of $1 each. The prospectus for the company begins by stating,

> The perfectness of this modern piece of mechanism has been attested to by the best engineers in this country, they having pronounced it one of the greatest inventions of the age. By its use air is compressed into reservoirs, whence it is conducted to engines that operate dynamos, thus generating electricity. As "necessity is the mother of invention," this wonderful discovery— enabling us to harness the waves of the mighty ocean—has come to the rescue of toiling humanity in the eleventh hour; for by its free and equal use man will be brought nearer to an equality in opportunity.[12]

It was big talk, but there's no suggestion in the historical record that the wave motor ever became something other than the model that Goldberg may have seen. But in California at the time, it must have seemed like wave power was on the verge of a breakthrough. Starved for power, during the decades sandwiched around the turn of the century the state was home to a burst of wave motor experimentation that is startling in its intensity and seriousness.

In San Francisco, isolated even from the water power available to its easterly neighbors, the city's promoters—who had much to gain from population increases—hungered for greater access to energy. Without it, the city could lose its spot atop the West Coast pecking order. Given the lack of cheap fuel or water power, having the Pacific Ocean sitting right there, uselessly pounding the city's coastline, was rather galling. In fact, in 1895 the *San Francisco Examiner* held a contest asking its readers, "What shall San Francisco do to acquire one-half million citizens?" This was the question of the day, upon which fortunes depended.

Out of thousands of responses, the contest's judges—including James Phelan, later mayor of the city and California senator—picked the following response: "Offer fifty thousand dollars 'bonus' to any inventor of a practical mechanism capable of commercially utilizing ocean 'wave power.'"[13] The suggestion had been submitted by one "Eureka Resurgam," a mixed Classical pseudonym meaning, "I have found it" (Eureka) in Greek and "I will rise again" (Resurgam) in Latin. The contest's selection was a powerful indication that San Francisco needed power—and that wave motors were considered a possible breakthrough technology that could get it.

But not everyone was buying what the wave motor guys were selling. "San Francisco is the home of the 'wave-motor,'" one skeptic wrote in the magazine *Machinery*. "One comes around, as I am informed from one to three times a year. The external swell always rolling in here works the wave-motor man into an ecstasy of invention and he persuades an opulent friend to invest in the scheme."[14]

Expecting such responses, wave motor proponents could snap back with the prediction of America's leading inventor: "Edison said only a few years since that electricity would be the future commercial power of the world. That is true," went one advertisement. "He also said the ocean waves would furnish the power of the future. That is also true."[15]

IF GOD INTENDED MAN TO FLY

As the century broke over America, there was another really tough technical problem receiving huge amounts of attention from small-time inventors across the land: airplanes. In fact, *McBride's* magazine wrote in 1903 that

> probably no other single subject, save that of navigation of air, have so much thought and energy been expended as upon the conservation and utilization of the power exerted upon our sea-coasts by the force of waves. And certainly since the days of alchemists and astrologers few themes of thought pursued for practical ends have resulted in so little reward to their students.[16]

Just two years before, Rear Admiral George Melville, chief engineer of the U.S. Navy, had issued a similar condemnation of air travel. Melville thundered,

> Outside of the proven impossible, there probably can be found no better example of the speculative tendency carrying man to the verge of the chimerical than in his attempts to imitate the birds, or of no field where so much incentive seed has been sewn with so little return as in the attempts of man to fly successfully through the air. Never, it would seem, has the human mind so persistently evaded the issue, begged the question, and 'wrangling resolutely upon the facts' insisted upon dreams being accepted as actual performance, as when there has been proclaimed time and again the proximate and perfect utility of the balloon or of the flying machine.[17]

However, doubters did not stop Californians from trying to fly and harness the oceans. Unlike us, they didn't know how the story of aviation and wave motors turned out. Both these crazy dreams were united in one southern Californian, Alva Reynolds, who was both an inventor of an aircraft he called the Man Angel and a wave motor in the first decade of the twentieth century.

The Man Angel was lighter-than-air and had paddles like a rowboat that the aviator could pump. Strange as it sounds, it flew the skies of Los Angeles to the delight of thousands.[18]

Reynolds's wave machine, designed with his brother George, received a glowing write-up in the *Los Angeles Herald*. The Reynolds was "perfect in detail" and would not break. "If any wave motor of which I have knowledge will be a success the Reynolds is the one," said a local engineer who also happened to be a director of the newly formed California Wave Motor Company. The paper editorialized that "the enormous value of such a motor to the world is almost beyond the grasp of the mortal mind."[19]

At face value, the article seems to indicate that the motor was nearly ready for action. A large drawing of the Reynolds wave motor sat beside the headline, "Will Generate Electricity By Ocean Waves." Another Rube Goldberg machine, it was to be built like a pier. There were vanes on the pylons that would spin when the waves came in, turning a crank that would pump seawater to a reservoir on shore, where it would run down through a standard hydroelectric generator.

How fortuitous that a new company would receive such a tremendous write-up in a major Los Angeles newspaper—and just a month after putting out a stock offering! Unfortunately, it appears to be the result of some underhanded dealings. The company's stock offering prospectus reveals that the managing editor of the *Los Angeles Herald*, Frank E. Wolfe, was actually a director of the company. The *Herald* article conveniently left off his name from the list of company directors printed in the paper.[20]

The company had a model of their invention built at 21st Street in Huntington Beach in 1909. They claimed to have "passed the point where we must stand over a working model and argue with crowds of skeptics as to whether or not our motors will work in the ocean." But then the trail goes cold. Certainly, the wave motor did not ever become anything close to a commercial success. By 1911 Reynolds had lowered his sights, filing a patent on a technology to protect pylons from barnacles and parlaying it into a new company, the Common Sense Pile Protector Company.[21]

Another wave motor inventor, Fred Starr, saw his dreams meet a more public demise in 1907. Just up the coast from the Reynolds at Huntington Beach, the Starr Wave Motor Company built a huge power plant off Redondo Beach. Starr, who had finished the hardwood interiors of railroad cars for twenty years, initially opened up shop just a block away from the old Wave Motor Air-Compressing Company offices. He built a small working model, then a larger one at Pier 2 in San Francisco. Then the earthquake of 1906 hit, the city was destroyed, and Starr left town for L.A.[22]

At first, all seemed to be going well. The papers carried photos, not just drawings, of the construction of the plant by the Los Angeles Wave Power and Electric Company. The entity pumped its stock offering in the selfsame papers, trying to gain the technological potential high ground. However, the question remained: Should wave motors be clumped with the great inventions of the era—airplanes, cars, the Westinghouse airbrake, electricity—or the flops of the past like alchemy, perpetual motion, and patent medicines?

One article in the magazine *Overland Monthly* was clear on its take. Journalist Burton Wallace wrote,

> Wonderful as are the wireless telegraph, the Bell telephone, and the Mergenthaler typesetting machine, which set civilization forward nearly a century within the past decade, there comes now a remarkable invention, made practical and put into operation for commercial use at Los Angeles. It is called the Starr Wave Motor.

The company played on the environmentally benign aspects of wave power. Notably, it did not create the smoke or soot associated with coal and oil burning. One December 1907 ad trumpeted these benefits:

> A SMOKELESS CITY. Many things have been written and said of the natural beauty and charms of Los Angeles. Some unpleasant experiences have been had and unkind things said about our smoke nuisance. . . . Some of our citizens are now so aroused that they declare Los Angeles must be the first model and ideal smokeless city of this country."[23]

Starr went on to declare that by December 1908, "Los Angeles will be a smokeless and sootless city, clean pure. It will be made so by all the power and heating plants being supplied with power and heat from the ocean waves by the Starr Wave Motor."

Things did not quite go according to plan, however. By October Starr had been forced out of the company and was recuperating in a mental hospital from a nervous breakdown. The company's secretary told the *Los Angeles Times* that the enterprise was broke. "It was impossible to sell any stock because the plant had not started and produced electricity as promised," he said.[24]

Eventually Starr wrested back control of the company, but it didn't matter. In February of 1909, a mere two weeks after Starr had lauded the company's prospects again, the $100,000 Redondo pier and motor sunk "like a lump of sugar when dropped into water."[25] And by May of 1909 one thousand shares of Starr motor company stock of "par value" of $1 each could be had for $0.65.[26]

The *Encyclopedia Americana*'s 1920 edition summed up the experience of the Starr wave motor and all the rest from the period: "The history of all other devices that have been tried is more or less similar," the wave motor entry declared, "and educated engineers have come to regard the wave motors as akin to the perpetual motion delusion."

The California wave motor story, then, is essentially and deeply one of failure. Unlike windmills or solar hot water heaters, wave motors proved technical failures. They generally didn't work at all or only worked for a short period of time. Still, other attempts at creating wave motors have been made throughout the years. More than a thousand patents exist for devices to convert wave power into usable energy. But the enthusiasm for wave motors that swept California in the two decades around the turn of the century has never been matched.

However, the prize has not gone away. Figuring out how to capture the ocean's awesome energy would still change everything, as the inventors at the time believed. Wave power could still help the world get free from what Duffy's prospectus called "coal and all its disagreeable accompaniments."[27]

Climate change and increased fears over national energy security have countries turning back toward local, renewable energy sources. The backwater research field of wave power has received a tremendous boost of interest in the last decade. Twelve wave power companies now have prototype machines ranging in size from 5.5 kilowatts all the way up to the 750-kilowatt Pelamis "snake."[28] Furthmore, in March 2010 the United Kingdom gave licenses to a series of companies to develop 700 megawatts of wave and tidal power off the coast of Scotland. The UK Energy Minister, the *Wall Street Journal* reported, "described the wave and tidal industry as a 'second industrial revolution.'"[29]

If the projects pan out, wave power companies will finally realize the California Wave Motor Company's vision for what working wave motors would mean to the world. As the California Wave Motor Company prospectus argued,

> Volumes might be spoke and written on the subject of what this means to the world, but the person of average intelligence can see the hand-writing on the wall if they have noted the progress of invention and development during the last thirty years. It means that a new field for investment and opportunity is opening to the world. That an epoch has been marked in the industrial development of this country. It means that the largest known source of natural energy has been tapped, and the power-house of old Neptune made available for mankind. It means that the electric current is to invade the markets of the coast countries and become a competitor with all other forms of fuel.[30]

But, as it's been for the last hundred years, no one is quite sure if the new wave machines purported to be floating along European coastlines within a few years will fare any better than their forebears along the California coast.

chapter 9

Compressed Air and Electricity

A ROUND THE TURN OF THE CENTURY, transmitting power was quite difficult. In order to move power, it needed to be touching something else. Energy markets were local, so people tried to tap local resources. For those who lived on the beach, the waves sure seemed like a good bet.[1]

The development of high-voltage power transmission, however, changed the dynamics of power generation and usage. The ability to transmit electricity made tapping local power sources less appealing. If one could send 100,000 horsepower from a coal plant on the edge of town, why bother with a small wave or sun motor out by the pier? Transmission reshaped where and how people could live and work. As much as the fossil-fuel plants themselves, the means for transmitting electrical power that could be used for any number of things transformed the American energy system. Power went regional, and in the process, a lot of marginal green technologies fell by the wayside.

The tremendous success of this electrical generation and transmission has erased a fundamental piece of its early history: To the engineers and entrepreneurs of the time, the rise of electricity as a way of transmitting power was far from assured. One of the most fascinating episodes in energy history is the brief but intense battle between compressed air and electricity as the way to send power long distances. In just 1877, when Sir William Siemens told the Iron and Steel Institute of Great Britain that "a copper rod three inches in diameter would be capable of transmitting 1000 horsepower, a distance of say, 30 miles," the statement "startled the audience considerably," and "a smile of incredulity was observed to play over the features of many of his hearers."[2]

An entire alternative power distribution system was even developed in France during the 1880s. If you were walking the streets of gay Paris during that time and you wanted to know the exact time, you would have looked to one of the tall, ornamented clocks that dotted the city. They looked quite like any of the clocks visitors would have known from other places, but with one key difference: All of them were kept in absolutely precise alignment, their minute hands all moving at nearly the same moment.

Inside the clock, a wondrous operation occurred each minute: "the pneumatic distribution of time."[3] A burst of air raised a tiny vertically oriented bellows. The rod atop the bellows tripped a lever and the lever engaged with a sixty-tooth wheel that moved exactly one tooth.[4] The wheel's movement took the minute hand of the clock with it, so the sum of all those steps was to make one minute become the next at more than 7,500 machines across the city.[5]

The bursts of air that set this mechanical ballet into motion traveled across the city in pipes hung from the roofs of the city's sewer mains from a building in east Paris on Rue St. Fargeau. Enterprising Parisians soon found that many other kinds of machinery could be run through the system—not to mention an entire early letter transmission system.[6] A metering system developed that allowed customers to be charged for the amount of air they consumed. Usage exploded across the city in the 1880s.

In 1890 a London engineering professor found more than 225 compressed air motors scattered across the city ranging in size from half a horsepower to fifty horsepower. *Figaro*, *Petit Journal*, and other publications used compressed air for printing. All kinds of artisans were on the network, too:

> They are also used in the workshops of carpenters, joiners, and cabinet-makers, of smiths, of umbrella-makers, of collar-makers, of bookbinders, and naturally in a great many places where sewing-machines are used, both by dressmakers, tailors, and shoemakers, and from the smallest to the largest scale. They find application also in all sorts of industrial work, with

confectioners, coffee-roasters, color-grinders, billiard-ball makers, in many departments of textile industry and other matters too numerous to mention.[7]

What had been a tiny operation in 1870 had become a big business by 1892. A large central power plant was built in an iron-and-glass building marked by its two brick chimneys on the banks of the Seine.[8] Inside the high-arched building, coal boilers powered twelve 2,000-horsepower air compressors. For every horsepower hour delivered at the end of the line, 3.3 pounds of coal had to be burned. Victor Popp, the Viennese inventor who devised the system, noted with pride that this fuel efficiency "is under all conditions superior to the best results obtained by electrical power transmission."[9] Indeed, even in 1900 power plants required 67,000 British thermal units of coal to generate a horsepower hour of electricity. That's at least seven pounds of coal.[10]

Compressed air was seen as a legitimate competitor for electricity for the transmission of power. In 1890 there were no long-distance power lines in use, but the entire city of Paris was undergirded with compressed air transmission lines. Popp's success gave tinkerers all over the world a new component for their dreams. Electricity, after all, had been used primarily for illumination, not industrial power. Induction motors—now two-thirds of those in use—had just been invented by Nikola Tesla, but they were not fine tuned. Electricity was not yet the universal power, and many people were still kind of frightened of it.

Meanwhile, compressed air could be used by almost anyone for almost anything. The same London engineer went to a roller coaster park and "found a large horizontal engine placed in a recess driving a dynamo and cells for the electric lighting of the whole building; a small vertical engine in another place worked the rotary pump, which actuated the 'cascade;' two or three large air-driven fans in wooden shafts served for ventilation; and lastly, a simple connection on a flexible pipe threw the air-pressure into beer-barrels as they were brought in." Thus, compressed air could seemingly do just about anything.[11]

The success of the technology and the hearty willingness of the people of the time to project improvements made compressed air a hot topic. In early 1891 the engineers of the Niagara Falls Power Company were trying to decide how to deliver 200,000 horsepower from the Falls to Buffalo, twenty-two miles away.[12] This may sound like a small feat now, but it was a massive and unprecedented project that the company, in retrospect and with deserved swagger, called, "the great step in the transition from mechanical power in industry to electrical power everywhere."[13]

But it seemed just as likely that it might have been compressed air everywhere. The Niagara Falls Power Company's engineers received much conflicting advice from the engineers of the day when they held an international competition for ideas on how to develop the falls and transmit its power. The submissions they received used alternating current, direct current, and compressed air. Victor Popp argued strenuously that compressed air should be used in the power project precisely because it could be "absolutely based on experience. Nothing must be new or untried in machinery, or even of exceptional dimensions."[14] Popp and his collaborator, A. Riedler of Berlin, put forth a proposal, based on their experience in Paris, showing how the system could work. Representatives of the Niagara Falls Power Company had even visited Popp in Paris to see the system for themselves in 1890. A consensus was building to use compressed air to transmit power from the Falls to Buffalo.[15]

But in the summer of 1891 a German team built a hundred-mile alternating current, high-voltage, three-phase transmission line from a hydro generator to Frankfurt. It had an efficiency of 75 percent. No compressed air system could come close. The original compressed air plan was tossed out and polyphase AC transmission continues to rule the grid to this day.[16]

The point of this history is not to suggest that electricity wasn't ultimately a better system. As engineers figured out how to run higher voltage power lines, the amount of power that electrical wires can transmit with little loss became astonishingly large. Modern transmission and distribution systems can carry hundreds of thousands of megawatts across whole states while losing less than 10 percent of the energy that was pumped into the system.[17] Imagining a hydraulic or pneumatic system that could work as well is difficult.

But however preordained electricity's preeminence may seem with the benefit of hindsight, people at the time weren't really sure if electrical transmission over long distances would work until the Niagara Falls power station proved it could. Even after that, many questions remained: Would it be safe? Would it last? Could their employees learn to work with it? How much would it cost?

Technology historian Eric Schatzberg has noted that many accounts of why one technology wins out over another depend too much on uncertain cost data about what technology was and would be cheaper. What he notes for the trolley industry was true for the transmission of power, too. Schatzberg related that

> although electricity ultimately proved cheaper than all alternatives in terms of costs per passenger mile, comparative costs do not provide a sufficient historical explanation for the success of electric traction. Reliable comparisons of the actual costs of competing systems were almost unknown, especially comparisons that accounted fully for capital costs and depreciation. Predictions of future costs of competing technologies were even more uncertain.[18]

No one knew how good electricity and compressed air could get as carriers of energy. Modern historians tend to see compressed air as a transitional technology between the gearing, shafts, and muscle power of the nineteenth century, but some contemporaries saw compressed air as newer than electricity. Conditioned with the techno-optimism of the time, they saw pneumatics as the next-generation power *after* electricity.

In fact, one compendium of technical achievements called compressed air "a new force which is coming into general use, and is regarded by some people as likely to become a rival to electricity."[19] An 1896 *New York Times* article forewarned, "Should compressed air prove to be the efficient and economical motive power of the future for street railroads, it will have the curious effect of superceding electricity before it is fairly out of the experimental stage." That was fine by the author because electricity required a "hideous tangle of overhead wires [that] has overspread cities like a cobweb." The displacement of electric rails and their infrastructure would have been "an inestimable blessing."[20]

Even in 1899 one board member for a New York elevated rail company made the case for compressed air. Several major railroad companies in other cities had told him "some wonderful things relative to the experiments with compressed air." Compressed air was cheaper per mile than steam or electricity and "was more effective." In response, he stated,

> This matter of electricity seems to me to be rather uncertain. No one seems to know so very much about it. We have 300 engineers on the road now to whom we pay $3 a day apiece. They know how to handle steam, and if put on electricity we may have to educate them all over again. With compressed air the handling is simple and not dangerous. . . . I think compressed air is the coming motive power, and so do many members of the Manhattan Board of Directors.[21]

But it did not happen. Electrical innovation had kicked into high gear and Americans were fascinated by this new power. They wanted their country's name to be written in bright lights—bright electric lights.[22] Although compressed air remained an important industrial tool in shops and factories around the world, it never regained its position as a competitor with electricity after the turn of the century.

Out in California, in July of 1895, Sacramento became the recipient of electric power transmitted down twenty miles from a generating station on the American River in the foothills of the Sierra Nevadas. The Niagara Falls project, noted above, is often considered the defining moment for electrical transmission, but the Folsom Dam and Powerhouse, built by jailbirds from Folsom County prison, sent electricity nearly as far as the Niagara-Buffalo connection—were completed a year earlier. The Folsom project was considered a monster success. It lit up 12,662 incandescent lamps, 582 arc lamps, provided power for 35 streetcars, and 537 horsepower worth of motors in breweries, mills, and shops.[23]

Though compressed air advocates stuck around, the short, intense battle between power transmission options was already being decided in favor of electricity. The impending success of the project brought out the entrepreneur in the post-gold California spirit. The *San Francisco Call* wrote in June 1895 that a new kind of "hustler" has arisen, and within the past three or four months he has been rapidly multiplying and filling the earth. He is the promoter of new electrical enterprises, and especially the promoter of schemes for long-distance transmission of electric power. The air of the whole Pacific Coast has all at once become filled with talk about setting up water wheels in lonely mountain places and making them give light and cheaply turn other wheels in towns miles away.[24]

Within five years, exactly such transmission lines had been hustled into place. Electric power transmission lines ran to Los Angeles, Bakersfield, Fresno, Stockton, Sacramento. Around the country, the same thing was happening. The great centralization of the American power system had begun. It would continue unabated for the next eighty years until the great energy rethink of the 1970s. The country's electric grid now extends 157,000 miles across the country, mostly in three large networks: the Eastern, Western, and Texas Interconnections.[25]

III.

What Might Have Been

chapter 10

The National Electric Transportation System That Almost Was

ON AUGUST 31, 1894, two young men rolled their new electric car onto what passed for a road in Philadelphia. It would have been hot and sticky outside, a Friday at the end of a long summer that had seen an intense heat wave suffocate the city for most of July. Piles of manure would have dotted the stones.

As the two men began their slow ride, people must have stared. Horses must have balked. It was almost undoubtedly the only car in the city. Credit for the first American electric vehicle is generally given to Boston's Philip W. Pratt for his lithe three hundred–pound tricycle, but this new vehicle was one of the very first automobiles in the world. Even eleven years later, only five hundred cars were registered in the city.[1]

Pedro Salom, a chemist, and Henry Morris, an inventor, had built their new ride in only two months. As much tank as carriage, the Electrobat, as they called it, weighed 4,400 pounds and was powered by an adapted ship motor. Its designers freely admitted that the vehicle was not designed for "an attractive appearance to a carriage builder's eye."[2] Instead, they built the vehicle rugged because they wanted it to stand up to the rough city roads—not at all smooth like the roads of today—and they happened to need a place to put 1,600 pounds of lead-acid batteries.

We're not sure where Messrs. Salom and Morris went that first day, but over the next few months they buzzed all over the city, from Salom's house in the newly fashionable area by the 49th Street station all the way across the Schuylkill into downtown and even into Fairmount Park, where just a few years later the horseless carriages of the young

71

and restless would become common enough to require an ordinance explicitly permitting them.[3]

The first version of the Electrobat, on which they glided through the streets of Philadelphia that fall and winter, looks like an uncovered wagon, complete with the spoked wheels—big ones in back, small ones up front. Two could comfortably sit atop the battery compartment, which housed the monster lead-acid cells, but it could have carried up to a dozen people. It gives the impression of a stagecoach missing both the horses and the coach, but it got the job done. It had a maximum range of fifty to one hundred miles and traveled hundreds of miles in its few months of testing, if Salom is to be believed.[4]

Its successor, the Electrobat 2, weighed closer to 1,800 pounds and packed a couple hundred pounds of batteries. It looked like a box on wheels, and a conductor sitting in the middle of the front of the car drove it with a steering stick. This automobile was the one that would propel Morris and Salom into history.

The week before Christmas of 1895 Salom showed up at the Franklin Institute in the Electrobat 2. Sessions at the Institute were like the TED talks of their day. The most exciting science was discussed. For example, a few weeks after Salom spoke, audience members were wowed by Roentgen rays — known to you and me as X-rays — that let humans see their skeletons right through their skin. Behind the stage, a screen hung onto which lantern pictures were projected. The Institute helped define the landscape of mechanical dreams. In this august setting, Salom delivered a talk on the subject of automobiles, most specifically his own and its advantages. "The subject of automobile vehicles is almost as old as the locomotive," he began, and proceeded to show images of hallucinations of automobiles from classics of literature: Homer had dreamed of "self-moved" tripods that were "instinct with spirit," and Milton had described the Chinese, who "drive / With sails and wind their canny wagons light."[5]

Then Salom launched into his adventures with the Electrobat. The car had made a fine showing at a car race sponsored by the *Chicago Tribune* the previous month. It didn't come close to finishing the race, but it was awarded a gold medal for handling and good looks.[6] Finally, he made his plea for electric vehicles in thirteen easy points. The electric car was safe and clean; it made no noise, vibration, or heat; and it seemed unlikely to cause environmental problems if produced en masse. In a way, Salom foresaw 1970s Los Angeles:

> All the gasoline motors we have seen belch forth from their exhaust pipe a continuous stream of partially unconsumed hydrocarbons in the form of thin smoke with a highly noxious odor. Imagine thousands of such vehicles on the streets, each offering up its column of smell as a sacrifice for having displaced the superannuated horse.

Each of his points was an attack on the gasoline-powered vehicle, which was something like a bomb on wheels. It was loud as a motorcycle at Sturgis, vibrated like a jackhammer, threw off enough heat to burn its passengers, and belched smoke. The competition between electric, gasoline, and steam-powered horseless carriages was real.

At the turn of the century each type of automobile had about a third of the market.[7] We can be sure that proponents of each method of propulsion—not to mention the "lovers of horseflesh"—had frothy-mouthed adherents who would have left nasty comments all over the Internet had such a medium existed. In 1898 one Philadelphia electric vehicle proponent explained, "The electric vehicle possesses so many advantages over those propelled by oil or gas motors that, except in special conditions, where extremely large distances have to be covered without charging facilities, there seems to be little, if any, doubt that the former type will be the one to fill the field."[8]

Many nineteenth-century heavyweights agreed, and the Philadelphia duo became a part of the most ambitious effort to create an integrated, nationwide, electric-powered transportation system that the world has ever seen. Morris and Salom's second Electrobat became the technological basis of the Electric Vehicle Company, the first corporate car concern in the world, the first cab company in New York, and, in the words of automotive historian John B. McRae, the "Monopoly that Missed."[9]

The sociotechnological battle between types of cars is one of the most intriguing technological battles in American history. From it, an entire new mobility system swept through the United States. It required new roads and ways of thinking about roads, changing land-use patterns enabled by electrification, fortunate fossil-fuel industry dynamics, particular quasi-Victorian mores, and even the latent anger of the urban working class. The changes were radical and they snowballed throughout the twentieth century, making Americans the most car-dependent people on earth. The system is so deeply entrenched now that developing alternatives seems far-fetched.[10]

But in the waning years of the nineteenth century, it was not apparent that gasoline-powered cars would dominate. A betting man, otherwise known as an investor, might have put his money on the continuance of two long-term trends: the increasing centralization and the electrification of the nation's transportation system. Mobility would be sold as a service, he might have wagered. Why would anyone want to buy a two thousand–pound hunk of metal powered by a controlled explosion of a substance known to be as dangerous as TNT? The same Philadelphia electric vehicle fan noted, "The last few years have seen the electric motor replace horses in our street cars, and those who have followed the conception and growth of the electric vehicle prophesy the same revolution in carriages, cabs, wagons, etc, in our cities."[11]

Thus, a certain kind of wealthy investor might have seen the electric vehicle as the perfect extension of the electric rail systems that blanketed every sizeable American city.

A NEW COMBINATION

In April 1899 William C. Whitney, a New York financier, walked out of his home on 5th Avenue, bound for Hartford, with a million dollars earmarked to jump-start the creation of a nationwide electric vehicle company.

Whitney was a robber baron, playboy, lover of fine horses, former Secretary of the Navy, and syndicate builder. He married well; his mansion featured a Marie Antoinette room. Whitney's henchmen purchased ceilings, walls, and chimneys from old European manses and reassembled them in his home. It was a pastiche of the best pre-Industrial crafts remade into a modern home across the street from Central Park. The library was hung with sumptuous velvet tapestries and was paneled in dark oak. Vast expanses of Persian rugs cushioned footfalls. In the cavernous hall that led outside hung an enormous portrait by the Flemish realist Van Dyck of Charles I astride a white horse. Charles I believed in the divine right of kings and fought Parliament twice to maintain his power. The message to all visitors to the Whitney residence was clear: from God to Charles I to W. C. Whitney.[12]

The million dollars was an enticement for Colonel Albert Pope, who was the country's leading bicycle maker, to tie up with the Electric Storage Battery Company (ESB). The ESB had bought out Morris and Salom's Electric Carriage and Wagon Company, which had successfully opened up a cab service in Manhattan with thirteen modified Electrobats. During April 1897, their first month at Broadway and West 39th, Morris and Salom happily reported to the Society of Western Engineers that they had served a thousand passengers and the small fleet collectively traveled two thousand miles across the city.[13] There was just one problem with the vehicles: They did not have the range of their gasoline competitors. Batteries, even our modern lithium ion ones, do not pack as much energy per pound or cubic volume as gasoline does.

That disadvantage could be mitigated with an efficient central station that would allow for fast battery swapping. During that year Isaac Rice and the ESB took over more active management of the enterprise. In particular, they asked George Herbert Condict to design a new way to swap batteries in and out of cabs quickly. Condict responded with an ingenious system that drew on his experience supervising a Manhattan streetcar line that used swappable batteries for power. The ESB constructed it in a converted skating rink at 1684 Broadway to service the rapidly growing fleet.[14]

When a cab drove into the station, technicians secured and centered it with hydraulic shoes. They then hitched the 1,300-pound battery tray, which ran underneath the cab, to a hydraulic piston that pulled out the whole thing and sat it on a table, where "an overhead crane plucked it from the table and deposited it in the charging room." They slotted in a new battery and off the cab went again into the wild Manhattan streets. Transportation historian David Kirsch, the most sensitive of contemporary historians of the company, called it "a marvel of modern mechanical engineering. For the first time, industrial practice was brought to bear on the age-old problem of transportation over city streets."[15]

The business caught Whitney's eye. In the early 1890s, while working Whitney's political connections, the Metropolitan Street Railway

Company received exclusive license to operate horsecars in Manhattan. Forgotten now, this mode of transport used horses to pull buslike vehicles along rails. The rails reduced friction, thus allowing equine teams to pull dozens of humans to and fro.

Although there were several options for mechanizing this form of transportation, Whitney's band of barons had gone with electricity, and it "was an integral component of their success."[16] He trusted electricity as a moneymaking enterprise and began to imagine a syndicate that could control all kinds of electrified mobility within and between cities. Electric trains called interurbans would run between local towns, trolleys would provide service along major routes, and the electric vehicles would serve any other intracity mobility needs. Urbanites wouldn't buy a car: They'd be able to go anywhere on one type or another of electrified transport.

The electric car of the late nineteenth century was a perfect vector for this vision. Slow by modern standards but faster than horses, they were quiet, clean, maneuverable, and relatively safe.[17] The electric vehicles of the time couldn't be driven long distances, but in 1899 that didn't really matter. In 1904 only 7 percent of the two million miles of American roads between cities were surfaced, and usually that surface was gravel, not asphalt or anything that could stand up to large amounts of traffic.[18] Who would build the roads that would support long-distance car travel? Besides, there were already railroad lines running between cities big and small. It seemed natural, then, to focus on transportation where the people needed it: inside the rapidly growing urban centers.

In a presentation to the Franklin Institute, a railroad engineer commented that all the "agitation for good roads" might be too late because "we may have reached an era of electric lines and bicycle paths rendering them unnecessary."[19] At the time it was possible to imagine an American transportation system that didn't include cars at all, let alone three hundred–horsepower versions that can go four hundred miles between fueling.

So Whitney got his boys together—A. B. Widener, Charles F. Ryan, and a host of other names that now adorn the big buildings of New York—and convinced them that there was money to be made displacing the old horse-drawn carriage with clean, noiseless electric cars. They would churn out thousands of electric vehicles, sending them to the big cities of the world—New York, Chicago, Mexico City, Paris—where they would seamlessly fit into the transportation web that crisscrossed the world's great human agglomerations. At the back end of all the mobility, there'd be the miracle of electricity, as represented by the central power plants of Edison Electric and New York Heat, Light, and Power, which Whitney and his band of scions of wealth also controlled.[20]

What five years earlier had been a simple two-man project in Philadelphia had morphed into a play to unify the transportation infrastructure of urban America into one great syndicate. What they needed was scale, and that's what Pope could provide.

He was the largest manufacturer of the product at the center of America's latest craze: cycling. Scholars estimate that some ten million bicycles were in use during the 1890s in a country with a population of seventy-five million.[21] During the real boom years around 1895, "Pope's factory at Hartford was running day and night with three 'gangs' (presumably shifts) of men, making 150,000 finished parts requiring 500,000 operations every 24 hours."[22] Though Pope's legacy is disputed, one scholar found "many continuities" between the manufacturing techniques of Pope and Ford, the great symbol of mass production.[23] At the very least, Pope would have seemed like a tremendous partner for building a global automobile manufacturing and service concern.

By 1898 the components of Pope's newly consolidated American Bicycle Company cranked out 800,000 bicycles.[24] They made their own tires and steel tube frames, and they assembled them in massive quantities. Pope's company had also been toying with an electric car concept that had yet to catch on.[25] If that was one reason Pope was receptive to Whitney's offer, the other was that Pope and Whitney, like many other businessmen of the late nineteenth century, believed that scale was the answer to any and all troubles in business and society. Creating an integrated transport company seemed likely to yield greater efficiency, according to the business maxims of the age. "All aggregations of capital, if rightly handled, tend toward the betterment of the public," Pope said in 1899, the year of the deal. "This is a doctrine which all of us have not yet quite comprehended, but the experiences of every passing year emphasize its truth. It seems to me that we are fairly entered upon a wonderful period of political and financial history."[26] Increasing centralization was tantamount to progress (with a capital P). Even American socialists thought "aggregations" were a good idea. Edward Bellamy's *Uncle Tom's Cabin*-level hit, 1887's *Looking Backward*, envisioned a society in the year 2000 in which every industry had been consolidated into one, big socialist enterprise.

In the novel, the wise man of Bellamy's future notes, "The fact remained that, as a means of producing wealth, capital had been proved efficient in proportion to its consolidation." To bring back the days of small business "would have involved returning to the days of stagecoaches."[27] Indeed, the only way to move forward was bigger and bigger companies—capitalism so super in its scale that it becomes socialism. The process of syndicate building "only needed to complete its logical evolution to open a golden future of humanity." Then, the government became "the final monopoly" as "the epoch of trusts had ended in The Great Trust."[28]

Thus, it was only logical that transportation would soon be monopolized by a few.

Pope and Whitney sealed the deal and each side of the transaction took half of the Electric Vehicle Company. As an enterprise for building and operating electric vehicles, it seemed to have all the right parts: the Electric Storage Battery Company and its patent on the lead-acid storage battery, the Pope manufacturing apparatus, Whitney's financial connections, and the central station service model developed by Condict.

BICYCLES AND ROADS, BARONS AND RAILS

As the Electric Vehicle Company (EVC) rounded into shape, there was a brief moment when it seemed that success might be at hand. The New York station was performing well and new offices began to operate around Boston, New Jersey, Chicago, and Newport.

But to say that the EVC was a grand disappointment would be an understatement. Within about a year problems began to appear. In New York the service remained profitable, but the other cities suffered from poor management and operations. The batteries were not properly cared for, nor were the drivers trained well. Led by the trade magazine *Horseless Age* and its "autoelectrophobe" editor, E. B. Ingersoll, the public started to call the company "The Lead Cab Trust." The regional operating companies were shut down in February 1901.

People began to suspect that Whitney and his financiers were merely trying to pull some stock swindle. That notion gained steam when the EVC turned patent troll and began brandishing the Selden patent, which it said covered all automobiles.[29] Automotive historians of the 1950s have tended to see the problems as simply the gurgling death cries of an electric vehicle industry being taken out by the insurgent gasoline-powered car; they see the death of the EVC as a demonstration of the technological impracticality of the battery-powered vehicle.[30] But contemporary historians like Gijs Mom and David Kirsch have taken the company more seriously. Kirsch sees the scheme, if not the actual company, as "the seed of an alternative transportation system for motorized road transport."[31]

This alternative transport scheme would have been an electrified adjunct to the existing rail and trolley lines. Urbanites could have gotten anywhere in the country on a combination of rails and electric cabs. It would have been far more energy efficient, but from a consumer perspective, it curbed autonomy.

That turned out to be very important because the company was swimming against a very important cultural trend: the massive popularity of the bicycle. It was the crazy popularity of the two-wheeled bike that laid the cultural, infrastructural, and legal groundwork for the privately owned, gasoline-powered vehicle's dominance. "Easily the greatest significance of the bicycle was the interference it ran for the automobile," wrote sociologist Sydney Aronson. "The bicycle did the dirty work for its mechanized successor in a variety of ways."[32]

Operationally, the more than six thousand American bicycle repair shops that existed in 1900 became the "logical repair place for the auto" and helped train a generation of mechanics and inventors who would go on to service and create new automobiles. Culturally, the bicycle pulled people off the rails. It got them used to thinking about traveling on their own, whenever they pleased. They turned mobility into a product, not a service. With railroads and horsecars and trolleys, a person paid simply for the ride, not the vehicle itself. Bicycles, however, were different. People owned the machine and could ride it on their own schedule, even late at night or out to where there were no other people. We take for granted how easy it is, even in a densely populated region, to drive to the middle of nowhere. That simply wasn't possible before the bicycle, and the sense that one could be truly alone was liberating.

What's more, people didn't have to worry that the robber baron in control of the trolley lines might decide to change the line or suspend service or raise rates. Each person was his (or her) own (wo)man, just the same as William Whitney.

Bicycles set up the expectation among urban Americans that transportation could be freewheeling and fun, selfish and impulsive. Like their rural cousins or parents, who could choose to ride their horses whenever they pleased, they could take the bike out for a spin at any time. These two-wheeled fun machines made a new activity accessible to the inner-city population: touring the countryside. Urban Americans in the increasingly coal-polluted cities of the Northeast discovered that they could purchase a bike for about a month's salary and ride it right out of town into an agrarian world that was much more like 1800 than 1900, at least as far as they were concerned.[33] It's no wonder that bicycles were all the rage! And they were getting cheaper by the minute.[34]

Revenue poured into the coffers of bicycle makers like Pope, who decided to reinvest some of the funds not just in traditional advertising but also in supporting civic lobbying groups dedicated to improving America's roads with taxpayer money. Pope founded the League of American Wheelmen and financially supported its agitation for what the group simply called "Good Roads." Sure, they engaged in silliness like racing and bicycle polo, but the group was also a potent, progressive social force. By 1896 the Wheelmen pressured sixteen states to appropriate money for the paving and improvement of roads.[35] By the turn of the century its ranks had swelled to one million members.

The bicycle adherents were paving the way for automobiles in a lot of subtle ways, too. Texas A&M historian Peter Hugill writes,

> The League of American Wheelmen not only agitated for good roads but also published touring maps and guides, erected road signs, and identified inns and hotels that provided appropriate accommodations for middle-class and upper-middle-class urban tourists who were seeking the pleasures of the American countryside. That level of organization and the emphasis on the

conveniences of touring formed the groundwork for the automobile owners when the automobile superceded the bicycle as the means to see the United States.[36]

This transformation of the road from a multi-use strip of dirt into a place for individual (and eventually high-speed) vehicles began with the bicyclists, too. The rules of the road, long established by tradition as much as law, began to transform. States began to require bicyclists to register bikes and equip them with bells. Riding on the sidewalk was prohibited. States passed statutes requiring people in an accident to exchange names and addresses. Pedestrians learned that people with wheeled vehicles could kill them, as a new column, "Death by Wheel," popped up in newspapers.[37] It was into this tradition—fun, individualistic, and dangerous—that Americans eventually slotted gasoline-powered cars, though at first they resented that the rich were having all the fun.

In the early years of the twentieth century, cars—fast cars—were becoming a must-have plaything among the children of the robber barons, particularly their sons. In fact, they were "arrogance of wealth" put on wheels and driven right through the center of what had been the public commons.[38] Early films of American cities show enormous variation in the use of the road. Hand carts, horses, horse-drawn carriages, and pedestrians all shared the same ribbons of dirt. Bicycles required some adjustments on behalf of other road occupants, but gasoline-powered cars were fast and heavy enough that they required (and still require) thoroughfares all to themselves, lest they kill or maim anything too slow to keep up. Gasoline-powered cars required changing the definition of a road from a community space into a car-only lane. Woodrow Wilson, then president of Princeton, not America, even told an audience that "Nothing has spread Socialistic feeling in this country more than the use of automobiles. To the countryman they are a picture of arrogance of wealth with all its independence and carelessness."[39]

But the Electric Vehicle Company and its vision of centralized transport could not have capitalized on the popular ferment because it was itself controlled by the fathers of young men speeding around the countryside, scaring the horses and occasionally killing the children of poor farmers. Even the chauffer of William Whitney's son, Harry Payne

Whitney, killed a French coffee peddler named Andrieux in 1907 with his powerful car.[40] If there was one thing that poor Americans liked less than the young men speeding around the countryside, it was the fathers who bankrolled those young men speeding around the countryside.

In that same year, the Electric Vehicle Company finally filed for bankruptcy, never having been able to provide the kind of integrated service that Whitney originally imagined. He died later that year.

As Kirsch has pointed out, the entire idea of a service-based, multi-vehicle, electrified system left the American mind with Whitney's would-be masterstroke. Kirsch wrote in his study of the company, *The Electric Vehicle and the Burden of Failure*, that

> the collapse of EVC—far from being an abject, irredeemable technological failure—can instead be seen as a turning point, not simply in the choice of automotive technology, but what is more important, in the choice of business concept that shaped the subsequent development of the American transportation system. The dynamic evolution of technological systems has left us with the decentralized, loosely coordinated, individual-based transportation system of today. In this sense, the fate of EVC took with it a particular alternative vision of personal mobility.[41]

The electric car, however, didn't die as quick a death. Hundreds of electric trucks served merchants admirably in the first decades of the century. Electric cars, in absolute numbers continued to be manufactured, but as we all know, the Model T was the iPhone of the consumer car market. Gasoline automobiles dominated from the teens onward.

The dynamics of the fossil-fuel industry had also changed remarkably from the founding of the Electric Vehicle Company in 1899 to 1913. In the late 1870s the world's then-largest oil field in the Petrolia region of Pennsylvania had given out. The boom-and-bust of the region led some to speculate that crude oil was much more rare than it turned out to be. Then, throughout the first decade of the twentieth century, huge oil finds in Texas and Oklahoma, beginning with Spindletop in 1901, began to allay fears that had plagued gasoline supporters. Fears of an "oil famine" quieted and slowly went away, further solidifying the hold that internal combustion engine vehicles had on the market.[42]

For bicyclists, all the hard work they'd done for automobiles would not be repaid. Once a relatively fast car reached critical mass, the slower alternatives were no longer available. Bicyclists, electric cars, and horse-carts all became hazards and had to be done away with. Roads had to be made the exclusive province of gasoline-powered cars—or vehicles that could closely mimic gasoline-powered cars' speed and maneuverability. Cities had to change ship to fit the needs of human beings in their new two thousand–pound exoskeletons.

A cartoon on the cover of the November 20, 1902, *Life* magazine shows a picture of the round earth covered with speeding automobiles trailed by billowing clouds of dust and exhaust. The humans are depicted falling off the globe and tumbling into nothingness. The title reads, "Who Owns It, Anyway?"[43] In 1905 a herd of sheep could dally near central Philadelphia; they'd have been scattered or killed a decade later. The motor car transformed what the road was. Once we went down the gas automobile path, the ride acquired tremendous momentum. Nonetheless, changing technological directions remains difficult. The car and its engine full of controlled explosions sent cities flying apart, each additional horsepower spreading them over ever-increasing distances. Now we may find a stray house or a strip mall somewhere far beyond any services or other houses, placed as haphazardly and without regard for physical geography as if it'd been blown there by a miner's roll of dynamite.

The electric rail lines that united every major American city in the early decades of the 1900s—and made robber barons like Whitney rich—were eventually ripped out of the ground and paved over. We now live in the cities we built for our cars, locked into a transportation system that is dependent on low oil prices in a world that no longer finds that prerequisite guaranteed.

The good news is that the last time Americans got fed up with a transportation system, they tossed it out when presented with an alternative they liked more, despite the wrenching changes it brought to communities and entire ways of life.

chapter 11

Solar Hot Water, Day and Night

O N ANY OLD EVENING in October 1929, one might have found Death Valley Scotty and his patron, Albert Mussey Johnson, coming back from a horseback ride among the rattlesnakes and scorpions to the sprawling Moorish villa that Johnson had built in the hottest place on earth, miles from nowhere. However, nights could still get a little nippy. And if one of them wanted a hot shower out in the middle of nowhere, he could have one courtesy of a new Day and Night Solar Hot Water Heater, which the duo had gotten installed earlier that month.[1]

Johnson was an insurance magnate and millionaire (even after the stock market crash that lit the Depression's fuse) who hailed from Chicago. Scotty was from Kentucky, but he had hung around long enough in Death Valley to become part of its natural fauna. He had been a cowboy in Buffalo Bill's Wild West Show and a confidence man for years. One of his first marks had been the city slicker, Johnson. Somehow, however, they became friends. No one is quite sure why.[2]

Neither, though, was known to be much of an environmentalist, or, as they might have been known then, conservationists. Why, then, install a solar hot water heater?

It wasn't that they did not have other sources of power available. They used the force of an underground spring to do mechanical work and run a generator, and they had diesel generators, too. During the same reconstruction of Scotty's Castle, as the compound became known, they also installed fuel tanks. They certainly could have decided to install some other kind of water heater. Day and Night itself offered a gas version of its heater.[3]

The fact of the matter is that the solar heater was probably the best option they had. Built from simple materials—concrete, copper loops painted black to absorb heat, and glass—the heater worked well, even if in the winter months, the water was warm, not hot. It could store 120 gallons of warm to hot water, as the name implies, all through the night.

The decision was a common one for consumers in southern California. The Day and Night Solar Heater Company was already generating $230,000 a year in revenue as far back as 1923, its owner William J. Bailey bragged to his Rotary group in Monrovia, California.[4] The *Los Angeles Times* even covered the rise of the company, noting in the subheadline, "Factory Forced to Move to Larger Quarters Twice as Demand Grows."[5]

Bailey had been trained as a mechanical engineer at the University of Michigan and then worked as a machinery designer at the Carnegie Steel Mills. Sent to California for the weather's salubrious effects on health, he decided to get into the solar hot water game in 1909, in concert with inventor Gilbert Cartter.[6] Bailey filed a patent application in April of that year and received it the following August.[7] They were in business.

But they weren't the only game in town. Clarence M. Kemp's Climax solar water heater had come into widespread use in the years following its development in 1891. Two Pasadena businessmen paid Kemp $250 in 1895 for the right to market the device in California. Within five years they had sold more than 1,600 in southern California alone at $25 a pop, at a time when Los Angeles had just 100,000 people. The hot water heaters were selling like hotcakes. Clearly, they were a good deal.[8] Ken Butti and John Perlin wrote in their wide-ranging treatment of the history of solar power, *A Golden Thread*, that "for an investment of $25, the average homeowner saved about $9 per year on coal—and more if artificial gas was used for water heating."[9]

These early hot water heaters were not sophisticated. The first, which predated the Climax, were simply bare metal tanks turned to face the sun. Kemp's invention incorporated the principles of the hot box. He put his water tanks inside a pine box fronted with glass. The glass trapped the heat from the sun like a car does when it's left in a Los Angeles parking lot.[10] The big problem was that if a person wanted to take a shower at night, she couldn't. That was the problem Bailey set out to solve—and he did.

His system used the natural tendency of hot water to rise—because it's lighter—and cold water to fall. The series of pipes he called "Sun Coils" circulated water through the pipes purely through the movement induced by solar heating. The hotter water eventually moved into a storage tank that was placed above the flat, glass-covered collector. The colder water fell and was circulated through the system again.[11]

If Scotty's Castle was and is one of the stranger places on earth, with its soaring turrets and swimming pool and handcarved furniture transplanted to what remains the middle of nowhere, the Day and Night hot water heater installation was a pretty standard job. Sadly, however, that sale was one of the last the Day and Night Solar Heater Company would make.

Day and Night's solar heaters had competed well against other hot water systems—usually just one's stove—that burned coal or gas manufactured from coal. With California far from the deep coal seams of the East, transportation costs made the fossil fuels expensive.[12] Beginning early in the century, though, gas started to get cheaper. One reason for this was that gas manufacturing plants began to be built increasingly larger. Through classic economies of scale, the gas that left the plants was cheaper than at smaller, less efficient plants. However, as big plants needed lots of capital and big markets, the gas industry began to consolidate. From 1902 to 1920 the Pacific Gas and Electric Company, which operates in northern California, doubled its gas output and miles of pipeline, and near Los Angeles, the Southern Counties Gas Company was completing a similar consolidation.[13]

At the same time, the discovery and subsequent exploitation of huge deposits of natural gas in southern California helped flip the energy economics around. Standard Oil, of Rockefeller fame, hit the gas well equivalent of a gusher in the Midway oil field near Bakersfield in 1909.[14] With the new supply, the gas companies expanded their pipeline networks to reach most of the major towns in California, mirroring the electrical grid in the state. By 1929 Pacific Lighting controlled a vast network of subterranean steel pushing natural gas around the southern half of the state at high pressure.[15]

The cheaper natural resource combined with better transmission made solar more expensive than gas. To make matters worse, the over-abundance of gas in the still-young state had the natural gas companies hunting for markets for their fuel. To oil companies, natural gas was a waste product of drilling for the real crude. In the early twentieth century 90 percent of the nation's oil-related natural gas was burned simply to get rid of it.[16] So the gas companies subsidized the up-front cost of the gas water heaters in hopes of growing the natural gas market.[17] Given how much gas was being wasted, finding ways to get consumers to use natural gas was, to the mind of Standard Oil's Frederick Hillman, "in the spirit of conservation."[18]

Bailey, a good businessman, saw this happening and introduced a new line of gas-powered hot water heaters. His company kept growing, but the once-burgeoning solar market didn't. The company sold a mere forty solar heaters in 1930.[19] The solar hot water business didn't die, though; it just moved to Florida. In that state, where gas wasn't as abundant as it has ever been or will ever be, the solar hot water heaters took off.

Miami was booming, nearly tripling its population from 1920 to 1925. H. M. "Bud" Carruthers showed up in the hot town with the Florida rights to Bailey's patent and a dream of building a solar hot water empire. With the California experience in hand, he set about creating the most successful solar industry of the first half of the twentieth century. "More than half the Miami population used solar heated water by 1941, and 80 percent of the new homes built in Miami between 1937 and 1941 were solar-equipped," wrote Butti and Perlin.[20]

By the late 1930s ten companies were competing for the solar hot water market. They would install perhaps as many as 100,000 units in Florida between 1936 and 1941. Furthermore, the newly created Federal Housing Authority financed the purchase of some solar units under a home improvement loan program.[21] This first federal solar incentive let Floridians buy a solar heater for installment payments of six bucks a month. It was no surprise that people bought units left and right.

Later solar hot water studies found that the technology remained cost competitive with fossil fuels into the 1970s and probably remains so today, but, nonetheless, the industry slowly died out after the war.[22]

It's a perfect example of how costs alone don't always determine people's choices in technologies.

As people looked to purchase new water heaters in the early 1950s, they tended to see old solar heaters that had been installed during the 1930s. Many of these had problems with corrosion in the water tanks. It was an easily avoidable problem in retrospect, but the solar installers didn't know that. A few burst tanks "badly tarnished solar energy's reputation of being a trouble-free way to heat water."[23] As solar's reputation was taking a hit, Florida Power and Light, like many utility companies, "mounted a major publicity drive for electrification." The grow-and-build plan for big power companies dominated their thinking throughout the middle of the century, leading them to push consumers to consume larger and larger amounts of energy.

This strategy often worked to push down the cost of a kilowatt-hour of electricity, as it did in Florida. By the early 1950s the price of that much energy was only four cents, down three cents from the previous couple of decades. With this tough new competitor, solar would have had to cut its costs to keep its price advantage. However, the use of copper, a major solar hot water material, was skyrocketing for other applications like electronics and infrastructure. Between 1930 and 1960 the use of copper in America nearly doubled.[24] Between 1938 and 1948 alone, its price actually did double, making the solar collectors more expensive at the moment when they needed to be cheaper.[25]

The costs associated with putting the installations on roofs were also annoyingly immune to cost reductions. Solar installers had the air of mom-and-pop builders. Each home was a specific job that had to be expertly evaluated and then worked on. Meanwhile, the electric water heater companies were building a box that did not need any special local knowledge. Add it all up and electric water heaters had largely closed the price advantage that solar water heaters had once enjoyed.

To make matters worse for the solar industry, "a new force, the large-scale builder-developer," showed up in Florida. These companies had a perverse set of incentives because they built homes before they sold them. Their only concern was driving down the up-front cost of construction. Electric water heaters fit their needs perfectly because they seemed cheaper, even if their cost over the life of the house could have been higher. "Developers almost always included electric hot water systems in new homes because of their low capital cost," wrote a historian of the period. "Monthly electricity bills were not their concern."[26]

With all that stacked against them, the U.S. domestic solar hot water industry slowly withered away. However, other countries picked up where American R&D had left off, notably Levi Yissar's work on new absorptive coatings in Israel. The Japanese market boomed, as did Turkey's and much of the European Union. But China became the big market. In 1991 the country had little solar heater manufacturing capacity. By 2005 thirty-five million Chinese families were using solar hot water heaters, with solar commanding a 12 percent market share in the country. In 2007 China had nearly 70 percent of the world's 2.2 billion square feet of installed collector capacity.[27] Chinese solar heater production outpaced Americans' by 160 times.[28]

Like so many other renewable energy industries, a field that the United States once dominated has moved on to greener pastures. A technology invented and improved in the United States is a dim memory here and a thriving industry elsewhere.

chapter 12

The Solar Home of the 1950s

ORLD WAR II was a time of deprivation in the United States. Coffee was rationed. Nylon was rationed. Fuel was rationed. The country made do, prodded by advertisements in *Life* and *Time* magazines, which promised an even more abundant nation as soon as the war ended. When it did, troops and war laborers began flooding home. For that first couple of years, with the memory of rationing fresh in their minds, a house that didn't need as much heating oil or gas seemed like a great investment to some Americans. They had seen the world's fuel supply chains torn apart, severed far more quickly than they had been built. The fragility of the world had been put on display. If the house was a machine for living in which some level of inputs yielded a certain level of service, it made sense to maximize service while minimizing investment. The solar house was like an efficient car.

George Keck built the popular notion of the solar house. The lore goes that Keck discovered the power of the sun while working on his "House of Tomorrow," which was built almost entirely from glass, for the World's Fair in Chicago in 1934. Though it was frigid outside and there was no artificial heat in the building, the workers had stripped off their coats and were "complaining of the heat." From that point forward, Keck spent decades working on harnessing solar energy to reduce heating costs in the buildings of Illinois.[1]

He was supported by a big business in America, the Libbey-Owens-Ford Glass Company. In 1935 the company came out with what it called Thermopane windows. They were composed of two sheets of glass with a layer of air sandwiched in between. The air acted as an insulator, keeping heat from escaping from the house while still allowing sunlight in. Although the Thermopane windows were only half as good at holding heat in as modern commercial windows with special glazings, the dual-paned windows were much better at keeping heat inside houses than their single-pane forebears.[2]

Though the Massachusetts Institute of Technology had been experimenting with more complex solar home designs that incorporated pumps and storage devices, Keck stuck with basic principles. His designs called for big Thermopane windows facing south with overhangs that helped keep out the high summer sun. In the early 1940s the success of his houses in cutting heating bills began to get attention in the Chicago daily newspapers. A local developer, Howard Sloan, who was living in a Keck home, began to promote an entire development in Glenville, Illinois. Sloan himself claimed to have saved 20 percent on his heating bills.[3]

As rationing ratcheted up American consciousness about energy, the fuel-saving solar homes began to look particularly interesting. Headlines in the major newspapers give a good indication of how optimistic this future looked: In 1944 the *Washington Post* proclaimed, "Sun May Aid Heating of Postwar Homes";[4] a *New York Times* article from 1945 noted projections of "Broad Demand for the 'Solar' Home";[5] and another *Times* article proclaimed "Purdue Experiments Show How Sun's Rays Can Help to Reduce Heating Cost at Home."[6]

Then, in 1947, Libbey-Owens-Ford Glass Company published a book with Simon and Schuster called *Your Solar House*, in which forty-nine architects designed homes appropriate for the forty-eight existing states plus the District of Columbia. In this, they were following the long traditions of vernacular architecture, building homes that suited prevailing weather conditions.[7] For instance, the Florida home had a glassed-in garden and a solar hot water heater. A two-story home was built largely from stone and tried to fit into the august "traditions of Delaware architecture." The Arkansas home was designed "using native materials" in a style "reminiscent of early American architecture"; it looks remarkably like any old subdivision home you might find, a nod to the architect's belief that "although startling scientific discoveries will have terrific impact on our way of living, it will take more energy than the split atom generates to change people's tastes and desires."[8]

To many, the solar home seemed the perfect match between the traditions of old and the science of the new age. *New York Times* critic Mary Roche wrote in 1945 that

> solar-house enthusiasts love to quote passages from Socrates and Xenophon and to cite building practices adopted by the Swiss 300 years ago and by the Chinese centuries before that, all by way of proving that solar heating is not a new-fangled idea. What has given it the aura of a machine-age miracle is the recent development of thermopane—the double plate glass pane with a metal-sealed dehydrated air space in between.

By 1947 whole subdivisions in the Chicago and New York suburbs were being built with solar home design principles. A 1948 *Washington Post* article exclaimed, "Solar Houses Win Approval Across Nation." That article went on to explain that in the first half of 1948, "more than 75 percent of the single family residences featured were of the solar type." Homebuilder magazines showed a similar trend, with 63 percent of the homes in those publications incorporating solar design.[9] The solar home looked to be on its way to a permanent place in the American building lexicon. Roche wrote that a new Keck house "will combine two principles widely acclaimed as inevitable in post-war home building—prefabricated construction and solar planning."

In the early 1950s President Harry Truman appointed a committee to examine the country's "materials resources." Headed by William Paley, CEO of CBS, they issued the Cold War–flavored report *Resources for Freedom*, and even the blue ribbon corporate panel foresaw a place for solar homes. "Efforts to date to harness solar energy economically are infinitesimal. It is time for aggressive research in the whole field of solar energy—an effort in which the United States could make an immense contribution to the welfare of the free world," they concluded.[10] Even without such an effort, but only with existing technology, they predicted that up to thirteen million solar homes could be built by 1975. Even to the corporate bosses of the day, using as much solar energy as possible to offset the country's fuel needs seemed like an excellent strategy. Americans, they believed, would think long term, choosing to pay a little more to build a home in exchange for the safety and peace of mind that came with needing less fuel. They expected the widespread deployment of solar energy by the mid-1970s.[11]

Their belief wasn't based on any far-reaching technological change. Evidence had accumulated over the course of the 1940s that fairly simple architecture that heavily incorporated glass and smart siting could reduce the costs of heating a home.

By the late 1940s prefabricated construction was taking off, and solar planning, as Roche noted, seemed poised to go along for the ride. Green's Ready-Built Homes, which had built a Keck house in its factory, estimated that prefab solar homes would "sell complete for between $6,000 and $8,500."[12] Prefab design reduced the cost to build and solar planning reduced the cost to own. What wasn't to like?

The scene seemed perfectly set for one of the major housing developers to adopt the new technology and use it to pitch to their lower-income clientele on the value of their homes. But solar planning required care and attention to detail. Without a skilled architect, solar homes could certainly go awry. One 1947 study found that for a particular solar home in Illinois, the heat lost at night through the large windows was greater than the heat captured by the glass during the day.[13] A modern architect looking back at the Libbey-Owens-Ford book also found much with which to take issue. Mainly, only Keck and Pietro Belluschi seemed to know what they were doing, and even they seemed to limit technical discussion of their ideas in the book. In 2008 Anthony Denzer, an architect at the University of Wyoming, wrote,

None of the designs explored more advanced issues of passive solar heating that should have interested a progressive architect in 1947. None addressed, even in general terms, the primary underlying objective of saving energy by reducing mechanical heating and cooling needs. None discussed how they had arrived at the amount of glass area relative to the volume behind it.[14]

As the 1950s began, developers did not want to deal with experts with their own sense of how a housing development should be planned and built. Not even the most basic features of solar planning—like south-facing windows—could be accommodated in the new subdevelopments. For example, the brothers who built the country's most famous suburb, Levittown, New York, chose to offset the aesthetic dullness of the Cape Cod homes by varying the houses' orientation to the street. Developers were simply not keen on anything that would add complexity to their projects. As environmental historian Adam Rome has noted about the scale of this buildout, every year from 1950 to 1970, an area the size of Rhode Island was bulldozed for suburban tract housing.[15] Thus, creating such massive amounts of homes required a certain strict dedication to speed and scale. In his blistering 1957 attack on developers, *The Crack in the Picture Window*, journalist John Keats wrote,

> The typical postwar development operator was a man who figured how many houses he could possibly cram onto a piece of land and have the local zoning board hold still for it. Then he whistled up the bulldozers to knock down all the trees, bat the lumps off the terrain, and level the ensuing desolation. Then up went the houses, one after another, all alike, and none of those built immediately after the war had any more floor space than a moderately-priced, two-bedroom apartment. A nine by twelve rug spread across the largest room wall to wall, and there was a sheet of plate glass in the living room wall. That, the builder said, was a picture window. The picture it framed was of the box across the treeless street.[16]

Such a builder was, therefore, not exactly interested in the vagaries of designing a passive solar house, even if he was interested in capitalizing on the "solar house" craze by sticking in a picture window.

Sadly, Keck's solar home prefabricator, Green's, demoted "the solar house to an experimental level" in order to "concentrate on a less expensive house." And even after the initial postwar building rush ended, the solar house did not make it back from that experimental level. It was eclipsed by a new model of futurity, which more neatly dovetailed with the desires of real estate developers: the all-electric home.

THE ALL-ELECTRIC DEVELOPMENT

The all-electric home sought to eliminate the elements, providing climate control at a level unthinkable before electricity became nearly free. Home builders realized that incorporating just a single energy source in each house—electricity—reduced their costs and gave them more flexibility with their floor plans, thus permitting them to offer "more house for the money." Not insulating or weatherizing homes, combined with using cheap building materials, allowed them to cut purchase prices, even if the house was far more expensive to operate.

The electric utilities and appliance manufacturers also actively courted their business, recognizing home builders as the key source of new business.[17] Utilities had to get bigger to make more money—and they believed that their "grow and build" strategy, in which people were encouraged to use as much electricity as possible, would deliver lower prices and greater market share. In fact, for the first six decades of the twentieth century, they were right. The utility industry was the site of an astonishing burst of innovation. Three things were working in their favor.

First, the thermal efficiency of power plants was improving rapidly. Engineers found ways to use more of the heat generated by burning coal into electricity. Edison's Chicago plant used nearly seven pounds of coal to produce a kilowatt-hour.[18] By 1965, however, new plants could generate the same electricity with just a pound.[19] There were no new technologies introduced during this period, only the kinds of metallurgical and process advances that allowed the utility operators to increase the pressure and temperature of their boilers, and new metals were developed that could contain the tremendous force used in the modern turbogenerator.

Second, for decades building bigger plants reduced the cost of each kilowatt-hour produced. In two big spurts in the 1920s and then after World War II, plants grew from just a megawatt or two into thousand-megawatt monsters. Millions of horsepower could be derived from a single power station. Although there was a technical reality underlying the love of gigantism, there was also engineering glory to be had by building the *biggest*.[20]

Third, the utilities pushed to have themselves declared natural monopolies in a geographic region so they could have ready access to money for building plants and fending off budding public power movements. The flipside of regulation was that utility profits were decided by utility boards that determined what their "fair rate of return" would be. Fairness, as you might expect, can be seen multiple ways, but in these golden days the slippery definition tended to aid the utilities because they had been able to make electrical power cheaper year after year.[21] Between the years of 1902 and 1930, say, the consumer price index doubled, but a 1930 kilowatt-hour cost the consumer 40 percent of what that energy power would have run in 1902.[22] The fall continued into the middle of the 1960s. The pursuit of size, efficiency, and a fair rate of return drove the industry.

Utility managers could do what they pleased professionally and aesthetically and still deliver higher profits and lower prices. What a deal! These men saw themselves as the spiritual descendants of Prometheus, the technical descendants of Edison, and the most benign kind of social reformers.[23] At the turn of the century electricity had been a mark of distinction for the upper crust; forty years later, it was the common technological property of everyone. As the Tennessee Valley Authority's David Lilienthal put it in 1944, electricity has "deep human importance, for this must be remembered: the quantity of electrical energy in the hands of people is a modern measure of the people's command over their resources and the best single measure of their productiveness, their opportunities for industrialization, their potentialities for the future." Lilienthal and others compared electricity to "a modern slave, working tirelessly for men."[24]

The only problem, really, was that consumers had to hold up their end of the bargain. That is to say, they had to keep buying more and more electricity. It was equally important that they buy the right kinds of appliances to smooth out the load drawn out of the electric system.

Here's why: Electric utilities have to build enough capacity to satisfy the largest possible load on the system, even if that's far higher than the average load. That means they end up sinking major money into building plants that might run only a small percentage of the time. Less dramatically, power plants work best when they can run consistently, so it's not ideal to run a plant to meet demand during the day and then switch it off at night. If you're a utility manager, you want the load to be as consistent and as close and safely possible to capacity throughout the day and the year.[25] The most picturesque example of this phenomenon comes from the early days of electricity. Back then, electric trolleys formed the dominant load, so companies built amusement parks at the ends of their lines to suck up nighttime load and also generate some extra traffic and revenue. This "load balancing" brought us Coney Island. By 1901 more than half of trolley companies operated at least one park with rollercoasters or underground tours or other rides. Historian David Nye has called them "a machine of illusions, the logical counterpart of the new industrial factory."[26]

Then, in the middle of the century more and more electric appliances that fit nicely into the grow-and-build strategy came onto the market: bigger, better refrigerators, televisions, washers, dryers, and curling irons. The modern idea of convenience was rounding into form. You name it, there was a new plug-in gadget being produced for the swelling ranks of the middle class.

In the classic story of energy in America, it's the amazing convenience and overall excellence of these products that drove electricity consumption. Demand growth averaged 6 or even 7 percent a year in the middle of the century, and at some regional utilities, growth was even faster, reaching the double digits.[27] Housewives just loved all their labor-saving devices! "The story of electrical progress is the record of the emancipation of womanhood—it has brought new golden hours of leisure to women and better living to millions of homes," General Electric proclaimed in 1936.[28]

And utilities were only too happy to provide the energy for this better world.

The story that historians have painted in recent years, however, is more complicated. Although items like electric irons and better refrigerators were undoubtedly useful and helped eliminate some of the drudgery from housework, historian Ruth Schwartz Cowan has convincingly shown that the more mechanized and electrified housewives actually did *more* housework than their less high-tech counterparts. How did that happen? From the 1920s onward the social standards for housekeeping transformed, driven by heavy advertising and marketing by home appliance manufacturers, who often also manufactured electrical equipment for utilities. For American housewives, social changes meant purchasing more products *and* doing more work. Cowan wrote,

> Not surprisingly, the changes that occurred were precisely the ones that would gladden the hearts and fatten the purses of the advertisers; fewer household servants meant a greater demand for labor and timesaving devices; more household tasks for women meant more and more specialized products that they would need to buy; more guilt and embarrassment about their failure to succeed at their work meant a greater likelihood that they would buy the products that were intended to minimize that failure.[29]

The transformation of household servants into electrical appliances was formalized when the Alabama Power Company created Reddy Kilowatt, an anthropomorphized lightning bolt with a lightbulb for a head and light sockets for ears, after which some three hundred companies utilized the image for their corporate spokesperson. He was the friendly face of private utilities' marketing campaigns from the middle of the 1920s into the 1960s, urging Americans to use more energy, thus enabling the companies to execute their grow-and-build strategies. "The most important elements that determine our loads are not those that happen, but those that we project—that we invent—in the broad sense of the term 'invention,'" head of American Electric power legend Phillip Sporn told his peers in 1964. "You have control over such loads: you invent them, and then you can make plans for the best manner of meeting them."[30]

Households were useful to utilities because their power needs matched up nicely with the corporate exigency of balancing the companies' electrical loads. One economic historian wrote,

> The household was recognized as a large and untapped market for electricity, capable of benefiting the electric companies in two ways. On a daily basis, household demands were steady and regular, which increased the total base load. On an hourly basis, the highest household demands came when industrial demands

were lightest, which evened out fluctuations in consumption. Consequently, the electric supply companies encouraged the production and sale of household equipment.[31]

In the 1950s the utilities paired Reddy Kilowatt with a new marketing campaign by the big appliance manufacturers to "Live Better Electrically." They hired famous actors like Ronald Reagan to promote a house in which everything was "at the Reddy."[32] Ads featuring the lightning bolt are embarrassingly strange, as Reddy, a near human, sings with the voice of a child: "I wash and dry your clothes / Play your radios / I can heat your coffee pot / I am always there / with lots of power to spare / because I'm Reddy Kilowatt." Then, he would climb into a light socket and deliver his tag line, "Remember: Just Plug in. I'm Reddy!"[33] Utility journals told their readers, "Sell or Die!" and "Sell—and Sell— and Sell."[34] Selling more electricity was not just good for business but also good for America in the long war against Communism: Electricity was, in the words of the Paley panel, a "resource for freedom."

The end result was that Americans used more electricity than anyone else and with far less of a sense of the value of conserving it. Whereas small electrical appliances helped increase load, the big market was in home heating and cooling. This interest united utilities and the burgeoning large-scale home construction industry. "We all hope that air conditioning is going to be the next thing in homebuilding," a leader of the National Association of Home Builders said in *House and Home* magazine, "but it won't be unless a great deal of changes are made."[35]

The changes that mattered were transformations in how homes were being built. The homes built in the 1950s were as close to antisolar homes as anyone can find. They were remarkable for their lack of attention to the climate of an area. Architects moaned about the loss of vernacular knowledge about the best ways to build houses in places as different as Georgia and Maine. One critic explicitly noted that Americans had "gone backward and unlearned things we used to know. We have built, in the South, little, low-ceilinged hotboxes without properly shaded roofs. We have built, in the North, houses with thin brick walls that are cold in winter and hot in summer."[36] Architects may not have known how to design a great solar home, but they did know it was a bad idea to build a house in Louisiana that was designed for Minnesota. Developers, though, were more concerned with following the trends that would sell homes.

So they built whatever homes they thought would sell and used energy input to replace architecture. The solar home's sense of economy and using less went out the window. Perversely, the more energy one was willing to use, the more house one could get for the money. The homes of the time became like stripped-down models of cars sold as door busters. Only what consumers found themselves buying wasn't just a cheap car with a big engine but rather an entire high-energy lifestyle.[37]

Perhaps there could have been consumer pushback on homebuilders who built houses that were wrong for an area and, therefore, expensive to heat and cool, but there was no equivalent to a "miles per gallon" sticker for homes. Some homeowners at Levittown, for example, never even saw the insides of their homes. They bought based on where the home was located within the development, not its thermal properties.[38] With long-term costs uncertain, homebuilders were most concerned with keeping their costs as low as possible so they could market cheap homes.

The problem of hot homes in the sunbelt built without proper insulation or shading was solved, for example, by air conditioning. There is no question that air conditioning is useful in warm climes, but the climactically inappropriate home designs of developers created part of the demand. "The use of air conditioning allowed homeowners to enjoy a new degree of comfort, but a goodly portion of the residential air-conditioning load simply replaced the comfort once provided—at little environmental cost—by good design," historian Adam Rome has found. In fact, air conditioning unit sales skyrocketed from a mere 43,000 in 1947 to 1,045,000 in 1953.[39]

For utilities, their success was actually a problem. Suddenly, they had a new peak load: the hottest day of the year when all the area's air conditioners were blasting cool air out of vents. The solution, as Philip Sporn could have told you, was obvious: If the demand isn't there, create it. The industry launched a major campaign to get consumers to adopt electric heaters. Unlike the amusement park load balancing of yore, there was nothing particularly romantic or interesting about an electric heater.[40]

If air conditioning was often an improvement, the electric heater could claim no such distinction. It's certainly no better than a furnace at providing energy service, and it's far less efficient. Power plants take fuel and burn it, generating heat that's transformed into the versatile energy form we call electricity. During the process, about two-thirds of the heat is lost. So if what you want is heat, it's better just to burn something or use the sun's free energy.[41] That's why, before the 1950s, only areas that had tons of cheap, hydroelectric power—thanks to federally backed programs in the Tennessee and Columbia River valleys—had adopted electric heat.[42]

Promoting electric in other areas, however, required considerable effort. Many utilities adopted rate structures that virtually guaranteed that only those with heavy usage required by electric heat would get the cheapest rates. If you used a ton of electricity, it was cheap. If you didn't use much, each kilowatt-hour was more expensive.[43]

General Electric and Westinghouse promoted the use of electric heat, too. Rome assigned them three interlinked reasons for promoting the inferior technology: "The two corporations sold electric heating units; the use of electric heat led to increased demand for a variety of household appliances, including air conditioners; and the overall growth of the power industry meant a growing market for electrical generating equipment, which the companies also manufactured."[44]

Thus, electric heat, despite being a remarkably inefficient use of fossil fuels, was a win-win for the manufacturers and utilities again.

The building industry liked electric heat because "all-electric" homes were cheaper and simpler to build because builders did not have to put in the infrastructure for natural gas. The utilities made sure that builders knew these benefits. And as the era of the built-in appliance rose, some equipment manufacturers increasingly sold to them. Home buyers at Levittown weren't just buying a house but also washing machines and televisions all rolled up into their mortgages![45] In later developments, electric heaters became inescapable, particularly in the Northeast.

Hiding just outside, or underneath, the advertisements for the all-electric house was another technological dream: harnessing the atom. Although the oft-quoted slogan that nuclear power would be "too cheap to meter" states the case too strongly, many scientists, engineers, and executives were convinced that eventually—in a time not far away in a galaxy very close to home—nuclear power would be cheaper than fossil fuels. As the 1970s wore on, the belief in the essential rightness of the build-and-grow strategy convinced American utility executives that they should continue to promote electricity usage, even when it began to become clear that nuclear power wouldn't be cheap and that concerns over the impacts of fossil fuel was changing their business.[46] It wasn't until the 1980s that utility regulators figured out that they had to "decouple" utility profits from the amount of electricity they sold before many were willing to stop promoting greater usage.[47]

But for the twenty-five years after the war, even solar researchers moved on. "Active" solar buildings that used pumps and more complex mechanisms for capturing and circulating solar heat became the scientific center of gravity. Little money for any kind of solar program was available, though.

The elegant and simple solar home, as conceived by George Keck inside the House of Tomorrow, had been forgotten. It wouldn't be until the late 1960s that a more radical group of architects and rethinkers would rediscover the sun's heat as a valuable energy input. In the meantime, millions of homes had been built without solar planning or climate considerations. An opportunity was lost to build a less energy-intensive stock of American houses, and we'll be dealing with the consequences of those decisions for decades to come.

chapter 13

The Solar Energy Research Institute

THE SOLAR ENERGY RESEARCH INSTITUTE (SERI) was born from the chaotic rush into alternative energy research that began in 1973 when the OPEC oil embargo shocked Americans out of their energy trance. The Solar Energy Research, Development, and Demonstration Act of 1974 called it into being, but for a variety of all-too-dull political reasons, the Institute's work did not begin until mid-1977 after Jimmy Carter took office.

Paul Rappaport was appointed the first head of the institute to be, in his own words, "the solar advocate for Mr. Carter." A respected scientist at RCA for twenty-eight years, Rappaport seemed like the ideal man to lead the nation's first and only government laboratory dedicated to what we'd now call green tech. "He is credited with many firsts in the development of solar cells," a June 1977 *New York Times* profile noted. "In his career at RCA, he has picked up 13 patents for electronics developments, has written extensively on technical subjects and has lectured all over the world. It is an impressive background."[1]

It seemed that SERI might be ready to handle its ambitious mission to advance "all of the solar energy technologies and all aspects of the process of moving a technology through the initial research stages to utilization in the commercial marketplace."[2] On May 3, 1978—designated as Sun Day by Congress—Carter visited the Institute, bestowing his blessing on it, promising to put a solar water heater on the White House, and announcing a surprise $100 million extra dollars for solar energy research. At other celebrations across the country, Sun Day was a very successful, if a bit crunchy, event. The Associated Press reported,

Typical of the activities across the country were those in Iowa. There were sunrise services and solar displays. There were songs about the sun and movies about the sun. After the ceremonies— which included a yoga exercise used by Eastern cultures to salute the sun—there was breakfast: granola, whole wheat muffins, and "sunrise soup," which explained one participant, contains orange juice, tomato juice, lemon juice, consommé and herbs.[3]

Some Sun Day organizers were not impressed with Carter's commitment to solar energy. One in particular, Denis Hayes, a senior researcher and author with the Worldwatch Institute, took potshots at Carter's energy plans from the gates of the White House. "The Carter Administration does not now have a solar policy," Hayes told the *Los Angeles Times*.[4] He sought a "fundamental change in policy," he told another paper, because solar had been "consistently discriminated against."[5] As we will see, he was right, no matter how the numbers were sliced.

Meanwhile, things were not nearly as groovy at SERI as the Sun Day festivities might have suggested. The days leading up to Carter's speech had been filled with lame hand-wringing in the press by the lab's staff because forecasts showed that it *might rain* during the president's visit. The worries revealed that the Institute was still as much a symbol as a working research laboratory that could rival a Los Alamos.

The rain did fall on Sun Day at SERI, and it was not an auspicious sign.[6]

Even in the best of times, SERI was not much to look at. The researchers operated out of rented space in an office park just off I-70, the highway that runs from Denver to the mountains and the great western desert beyond. Stubby office buildings clustered around small concrete ponds. Inside, the light was perfect fluorescent yellow. SERI looked more like a place where people made telemarketing calls or provided inside sales for a trucking company than the leading edge of a revolutionary solar movement. Although they had been promised a big, new campus with better research facilities, the designs were firmly stuck on the drawing board with no funds appropriated to build it. Unlike the big DOE labs, the Institute did not have a storied history, or Defense Department money.

SERI struggled to fulfill its promise as the largest group of re-searchers ever assembled to study renewable energy, and this occurred for four main reasons. First, the Institute was not given quite the level of support that it needed to pursue its wide-ranging vision. Its fiscal year 1979 budget was just $25 million, about half of what a National Acad-emy of Sciences report sketching out the tasks for a prospective insti-tute had recommended.[7]

The researchers had an enormous amount of ground to make up in comparison to other energy sources, as very little R&D had gone into making solar technologies feasible. In the fiscal year 1973 the lead solar agency in the land, the National Science Foundation, received less than $4 million for all of its research on solar technology.[8] From then on, al-though solar research budgets did grow, they never matched the big budget fossil fuel and nuclear technology R&D coffers. In 1979, how-ever, the federal solar R&D budget had grown to $484 million, a 12,100 percent increase in just six years. That year, all types of renewable en-ergy received total government funding and incentives of $1.36 billion.[9]

But that was still a miniscule amount in comparison to the federal budget as a whole, which broke half a trillion dollars for the first time that year. Even compared with other energy R&D projects, solar energy funding wasn't generous. The Synfuels program, which aimed to demonstrate the conversion of coal to liquid fuels, pulled in $4 billion from 1970 to 1984. Breeder nuclear reactors, zero of which operate in the United States, picked up an estimated $16 billion between 1968 and 1985.[10] What's more, the fossil fuels industries have received tremen-dous tax subsidies through the decades. The oil industry alone received $300 billion in tax incentives from 1953 to 2003.[11]

Second, SERI's research power was further diluted by the establish-ment of four Regional Solar Energy Centers, which came into being as a political compromise between the Carter Administration and Con-gress. Landing SERI was considered quite a pork-barrel prize for the delegation that could deliver its jobs, popularity, and prestige to its state. Nineteen states vied for the new lab, but when Golden, Colorado, was selected as the future home of the Institute, other states' representa-tives started lobbying for a more regional research approach. They got their wish with the Regional Solar Energy Centers. The final arrange-ment, however, was far from perfect. The Centers did not have a clear and defined purpose but instead siphoned off funds and focus from SERI's work. A close observer called the "birth pains" of SERI "a classical example of what happens when political expediency overtakes a basically sound idea."[12]

Third, the list of technologies SERI was supposed to tackle was long: solar heating and cooling of buildings, agricultural and industrial process heat applications, solar thermal electric generation, photovoltaic technology, wind energy conversion, bioconversion, ocean thermal energy conversion, and low-head hydroelectric power. The science and engineering underpinning each area was different, with only some crosscutting research possible. For instance, an expert in photovoltaics was not an expert in wind who was also not an expert in biofuels. The analyses turned out by SERI in the early years seem scattershot, lacking the direction expected of it.

Finally, Rappaport himself was not the right man for the job at that point in his career. Before heading out to SERI, the *New York Times* interviewed him at his office in Princeton, New Jersey. He seemed to view the job less as the challenge of a lifetime and more as a fun adventure in the wild West. "I think I'll live longer because of this," Rappaport said. "I think my blood will flow faster, and I'll be excited about it." He joked that the lab, located in the same city as Coors, would have three taps—hot water, cold water, and Coors—and pondered the leisure pursuits the West might offer. "When I get out there, I'm going to learn to ski. I might even take up horseback riding," he told the *Times*.[13] These were not exactly the words of a man dedicated to driving solar energy into the mainstream. As a result, malaise came over SERI that was noticeable even from the outside. The codirector of the Congressional Office of Technology Assessment report on solar power and later SERI employee, Henry Kelly, told *Science* magazine that the Institute had "enormous potential" under Rappaport, but "never lived up to it."[14]

The list of the Institute's problems grew. Its director of technology development resigned in disgust, telling *Physics Today* that "I have often found it hard to find a true supporter or dedicated advocate of SERI in the places that count."[15] The august publication *Science News* called the Institute "riddled by criticism and hobbled by chaos."[16] By the middle of 1979 the need for a change of leadership at SERI had become clear.

Shortly after the unveiling of a White House solar hot water heater, Paul Rappaport was rather unceremoniously fired.

His replacement was an altogether different kind of person: a gaunt activist with a deep, abiding passion for solar energy. They chose none other than the thirty-four-year-old Denis Hayes. The same man who, a year earlier, had stood outside the White House gates criticizing Carter and calling for a fundamentally new energy policy was now suddenly standing with the president's secretary of energy, James Schlesinger, being introduced as the head of the government's most prominent renewable energy initiative. "To some, it seemed as if the deserving were at least inheriting the Earth," wrote *The New York Times*'s Anthony Parisi.[17] One of Hayes's colleagues said to "expect fireworks" once Hayes assumed his role because "you're not going to see a quiet Denis Hayes."[18]

THE LONE ENVIRONMENTALIST

Hayes was a remarkable and incredibly unusual choice. At thirty-five, he was the youngest director of a federal laboratory ever. He did not hold an advanced degree, though he eventually completed a JD at Stanford. More intriguingly, he was an activist, not an engineer, scientist, or bureaucrat. Hayes helped found Earth Day in 1970 and became an influential and committed advocate for solar power, representing a key link between the environmental movement and alternative energy.[19]

Gus Speth, chairman of the Council on Environmental Quality, said he "couldn't think of a better person." Henry Kelly of the Congressional Office of Technology Assessment, called him "an interesting gamble."[20] Although his selection was unusual for solar energy, there was a long line of advocate-managers at the Atomic Energy Commission's national laboratories as well as in fossil-fuel regulatory bodies.[21] Hayes, therefore, gave solar someone who could match the persuasiveness of the leaders of the fossil fuel and nuclear camps. For years solar researchers had languished at tiny outposts on the edge of science. Their programs were barely funded, and their ideas were downplayed or outright mocked. Though some critics have argued that there was a "government takeover" of solar energy, the truth was that without government support, solar energy was unlikely to make its way into American lives.[22] Hayes's selection gave solar advocates one of their own arguing for their positions in the highest realm of government. After all, he had been a selection of James Schlesinger, the first secretary of energy, himself.

Schlesinger was not a radical man. A Republican with a pipe-smoking clubbiness about him, he had headed the Central Intelligence Agency before moving into energy. Thus, the appointment of Hayes seems downright mystifying. Hayes explained it like this:[23] Shortly after being appointed by Jimmy Carter, Schlesinger paid a trip to the oil-producing nations of the Middle East. The 1970s were a tense time for U.S.–Middle Eastern relations. In the wake of the Yom Kippur War between Israel and its Arab neighbors, the Organization of Petroleum Exporting Countries exercised their economic might, led by Saudi Arabia's oil minister, Ahmed Zaki Yamani. The OPEC actions dramatically raised the price of oil for Americans and touched off the energy crisis of the 1970s.

As Schlesinger awaited an audience with Yamani, he was seated in a rather dull room with few magazines or newspapers. A key exception, though, was a paperbook book with a sun on the cover. It was *Rays of Hope: The Transition to a Post-Petroleum World*, a book on renewable energy that Hayes had authored that year.

As it just so happened, Hayes had interviewed Yamani and, to thank him, had sent a copy of the work with a rather florid inscription (something that began like, "To my dear friend Ahmed," as Hayes recalled recently). Picking up the book and reading this chummy note to the Arab power player, Schlesinger turned to his friend James Bishop and wondered aloud, "Who the fuck is Denis Hayes?"[24] Bishop, who had been *Newsweek*'s DC bureau chief, happened to know and like Hayes, so good things ensued.

Still, Schlesinger asked a good question. Hayes is an unusual environmentalist. Though he loved blue jeans and sitting cross-legged as much as the next guy, he was not a trust fund vagabond or an urbanite who was ignorant of the chunks of the country covered with industrial infrastructure. He grew up in Camas, Washington, a working-class town in the southwest corner of the state along the banks of the Columbia River. His family never lived more than twelve blocks from the paper mill that dominated the small town. Their home wasn't extravagant, just solid brick and comfortable, with three bedrooms and a squat detached garage.[25]

Portland, Oregon's free spirit might have been forty minutes to the southwest, but Camas was a mill town, not a suburb. The local high school mascot was the Papermaker, and 75 percent of the city's tax receipts in the 1960s came from Crown Zellerbach's towering mill. Until the mid-60s, the mill's management and workers worked without too much discord. The local union was strong and run by a surprising mix of old southern Europeans, Greeks mostly, who'd been strikebreakers in the early twentieth century.[26]

The town had a complex relationship with the mill. It was one of the first mills that used the chemical intensive "Kraft" process in the Northwest. The paper got whiter, but the odor got much worse. Sulfur that was cooked out of the trees and added to the slurry during the papermaking process produced a sickly sweet smell that permeated the entire region.

Some local residents were furious at the changes to the new process. A local hotelier even won a lawsuit in the early 1930s for damages he sustained as a result of the Kraft process.[27] But the town's relationship to the mill was more complex than it might have seemed. One newcomer who moved to the town during Hayes's high school years summed up what many seemed to feel: "The attitude was, 'Well, that's the smell of money.' Because, as long as the mill was working and paying salaries and taxes, it is a good thing for Camas. There would have been no town without the mill."[28]

Camas was practically defined by its smell. At the local county fair, the town had a little booth to answer questions about the awful odor and try to downplay concerns about it. But people believed their noses. The smell from the mill could be overwhelming even miles away if it caught an unlucky wind. In town everyone was used to it, but the smell clung to their hair and clothes. If a couple went to a bar in Portland and pulled out a wallet or opened a purse to pay for a drink, the smell came stalking into the room.[29]

The environmental damage that the mill did was real and noticeable, Hayes recalled, but so were the economic imperatives that drove it. Hayes related,

Growing up, this simply seemed to be fate. Paper mills produced acidic fumes. That was a natural part of the industrial process to free up the cellulose from the lignin in the wood so that the fibers could be made into paper. All paper mills stank. Society needed paper; Camas needed jobs. The smell was "the smell of prosperity." As I grew older and learned a little bit about science and economics, I understood that "fate" is merely the sum of a large number of decisions made by people in authority who were trying to minimize their costs and maximize their profits.[30]

He went on to explain that he learned that if environmentalists wanted to clean up that one mill, "we would have to clean up the whole industry."[31] It was a lesson that would stick with him: Something as specific as the distinctive aroma of his hometown actually had national causes and implications.

After graduating from Camas High, he got an associate's degree at the local junior college, Clark College. He is by far the most high-profile alumnus the school has ever produced. Then he took the unlikely step of gaining admission to Stanford, where he became a powerful political leader, winning the student body presidency in that tumultuous year, 1968.

Two years later Senator Gaylord Nelson appointed Hayes one of the organizers of Earth Day and, quite suddenly, he became one of the more well-known environmental leaders in the country.[32] He had gained the national platform that he believed was necessary to effect real change. For the next eight years he wrote extensively, worked for environmental organizations, and founded the Solar Lobby in Washington, DC.

Then, after Schlesinger returned from the Middle East, he asked Bishop to arrange a lunch between Hayes and himself. "To our mutual astonishment, we quite liked each other," Hayes remembered.

Rappaport continued to struggle at SERI, as solar energy gained increasing prominence. Eventually, Schlesinger picked Hayes to lead the Institute. No other director of SERI or the National Renewable Energy Lab has ever come close to matching Hayes's record as an impassioned advocate for solar energy as both an idea and a set of technologies.

With a new leader at the helm and the 1979 energy shock pushing energy back up the political agenda, SERI morale began to return. In the fiscal year 1980 they had $131 million and a plan to spend it: They were going to systematically drive down the cost of the major renewable energy technologies. By making solar power cheaper, they would transform the relationship between society and its energy sources. Hayes recollected,

In 1980, if you looked at what was going on with solar energy and what was going on with computers, you could be pretty confident that America was facing a revolution but it wasn't a computer revolution. There was no Microsoft. There was nothing. Desktop computers could be used for advanced typewriting, some accounting and playing games, whereas on the other hand, on the energy field, we had a huge national enterprise with this research going on in multiple laboratories.[33]

The organization's institutional plan from fiscal year 1981 looked five years ahead and reflects Hayes's hopes and priorities. Photovoltaics got $38 million, more than twice as much money as any other technology, and its budget was predicted to keep growing. Solar thermal power research also received more than $10 million. Wind and biomass, though they received substantial sums, were not projected to get much more money. The big bet was on photovoltaics (PV), particularly what is known as "thin-film" technology, which uses less and cheaper material than traditional silicon cells.[34] "We had put in place a program to drive down the cost in a calm methodical fashion year and year after year not dissimilar to the one used in computer chips," Hayes said.[35]

In just the four years between 1975 and 1979, the cost of photovoltaic modules had dropped by a factor of three as money poured into the field and production increased.[36] The government guaranteed that they would purchase photovoltaic modules, which provided an indirect incentive for private companies to scale up and drive down the unit cost of PV.[37] Like so many technologies before them, photovoltaics appeared poised to ride a learning curve to mass adoption. "I was really convinced that we could do this thing. That we would drive these things down learning curves and get efficiencies of scale," Hayes said. "We were really going to foment something."[38]

They had another ambitious plan in the works, too. Congressman Richard Ottinger, who might be the most stalwart champion of green tech ever to pass through the legislative branch, asked the Deputy Secretary of Energy, John Sawhill, to create an "in-depth solar/conservation study." Drawing on the work of Lawrence Berkeley National Laboratory's Art Rosenfeld, Princeton's Robert Williams, the University of Michigan's Marc Ross, and SERI's Henry Kelly, the report sketched out an alternative vision for the American energy system that its authors felt would be cheaper and less environmentally destructive.

In fact, the takeaway message for utilities from the report was, in Rosenfeld's estimation, "Be wary before you invest prematurely in 50 GW of new plants (at $1–$2 billion each), the need for which is many years off."[39] The report was a direct challenge to the future that the energy industry said was inevitably on its way. The report said America could substantially cut its fuel usage while still maintaining economic growth by increasing energy efficiency and the use of solar energy.[40]

THE REAGANITES ARE COMING!

Politics, however, would intervene before Hayes's team had a chance to test their optimism. Jimmy Carter was crushed in the November election by former General Electric spokesman, Ronald Reagan.

Before the election, Hayes had been buoyed by a series of radio addresses that Reagan gave in which he promoted decentralized energy sources. It made sense to Hayes, too. Going off the grid is a radically conservative position in some ways, smelling as it does of self-reliance and Jeffersonian republicanism. But Hayes and SERI were in for a nasty reality check. As it turned out, the speeches had been written by a far-right Libertarian, John McClaughry, who envisioned a small-scale, rural democracy growing up in New England. He echoed many of those thoughts in the 1990 libertarian tract he coauthored, *The Vermont Papers: Recreating Democracy on a Human Scale*. The *Los Angeles Times* called the book "*The Small is Beautiful* of politics."[41] McClaughry wrote, "We do not feel Vermont will be able to work toward the strong network of small-scale local energy sources it needs until the political control over energy is decentralized."[42]

Reagan's politics were not McClaughry's, however. His transition team not only immediately went after SERI, but they also suggested closing the entire Department of Energy while maintaining nuclear research support structures in its place.[43] When that plan floundered under Congressional attacks, Reagan appointed former dentist and unabashed nuclear proponent, Allan Edwards, to head the Department of Energy. Edwards made it clear that "a vote for President Reagan was a vote for a nuclear future." He quickly proposed halving the SERI budget and cutting overall solar spending by 60 percent. In particular, those technologies closest to commercialization were the ones that would receive the least support.[44]

Programs that had just begun, like durability testing for solar collector materials and better standards for solar water heaters, were eliminated. By all accounts, the Reagan administration's attitude toward solar energy R&D had a "profound and mostly negative" impact on solar energy programs in the United States.[45]

Unlike nuclear power, which had survived several administrations with much of its funding largely intact, solar energy was not able to withstand the political change that blew in with Reagan. His administration began a large-scale rollback of Carter's solar initiatives, choosing to starve them of funds even if they didn't outright reject them. The Solar Energy Research Institute lost half its cash. Equally important, it was clear that solar energy was no longer what economic historian Steve Cohn calls an "official technology" anointed by government as worth pursuing. Investors and entrepreneurs realized that it might be time to get out of the solar game.

SERI researchers were devastated. In the early months of 1981, shortly after Reagan took office, Hayes arrived early at DOE headquarters for a meeting with the acting assistant secretary for conservation and renewable energy, Frank DeGeorge. As he walked the halls, checking in with friends and trying to gauge the sentiment of the staff under the new administration, a buddy ran up to him and asked, "Has Frank lowered the boom on you yet?"[46]

The boom, as it turned out, was that DeGeorge was going to suppress the publication of a million-dollar solar conservation report that

Ottinger had requested. So Hayes did the logical thing and ran out of the building. He called his secretary and told her to tell DeGeorge that he'd come down with the stomach flu and wouldn't be able to have the meeting. Then he called his lieutenant Henry Kelly and "told him that we needed to spend the next twelve hours Xeroxing everything that he had and mailing it out to a whole bunch of distinguished reviewers," Hayes said. "He and Carl Gawell [another author] stayed up all night copying and getting the reports in the mail."

The next day, when Hayes got the call from DeGeorge, he feigned surprise: "'Oh my god!' I said. We've mailed it out to fifty reviewers."[47]

The Reagan administration was not pleased. "His transition team was horrified by our draft report," Rosenfeld recalled.[48]

SERI was allotted no funds to publish the report. Ottinger held hearings on the suppressed report and entered it into the Congressional Record.[49] By that point, however, Hayes knew that the clock was ticking on his tenure in Golden. On June 23, the summer solstice of 1981, he was asked for his resignation. He composed an angry editorial to the *New York Times* imploring prosolar Americans to insist that solar energy "not be discriminated against." Congress restored some of the DOE's solar budget, but, as Hayes predicted, Reagan's ascendancy saw "the Federal solar program quietly eclipsed."[50]

The brief but grand solar experiment of the 1970s was over, and more than twenty-five years would pass before renewable energy funding would reach the levels it had enjoyed before. Solar electricity got cheaper in the intervening years, but the idea that the nation's energy system would get a total overhaul ebbed away. Solar still contributes miniscule amounts of electricity to the nation's grid, though the solar heat that warms every home to a greater or lesser degree goes uncounted in the official statistics.

Hayes went back to school, finished his law degree, and kept working as an activist. The years have only solidified his reputation as one of the most important and interesting environmental leaders of his time. Looking back on Hayes's career, a 1999 *Time* magazine profile gave Hayes a glowing evaluation. In it, David Jackson wrote,

> Where do you go when you need someone to rally 200 million people? An ex-President, perhaps, or a former dictator?

Whenever the environmental movement needs someone to gather the troops worldwide, it turns to a tall, understated activist who rides his bicycle to work, wears flannel shirts and has a unique ability to herd the masses toward a common goal. His name is Denis Hayes, but you can call him Mr. Earth Day.

But the man who could rally 200 million people couldn't overcome even the handful in Reagan's administration who were hell-bent on removing solar energy from the national energy debate. Perhaps no one could have.

"It was the best job I ever had; it was my one shot to reshape the world," Hayes reminisced. "If Carter had been reelected and we had four more years, I think we would have been unstoppable."

In the context of solar energy, the historian Frank Laird argues that "substantial enduring changes in policy require changes in the institutionalized ideas that influence policy, which can mean either changing ideas within an institution or changing which institution controls some policy." The Solar Energy Research Institute was formed to be one of those new institutions for changing the government, but the truth is that it never had a shot. Even under solar-friendly Carter, Denis Hayes and the researchers who worked for him were never taken seriously at the highest levels of government. And when Reagan came to power, they were pushed completely to the political periphery for decades.[51]

In 1990 the Solar Energy Research Institute was renamed the National Renewable Energy Laboratory (NREL). The name and formal recognition as a national laboratory were new, but its researchers still were renters in those same yellowing office buildings off I-70. The dream, stretching back to 1978, of a facility that would be "a national showpiece of innovative solar energy design and energy conservation techniques" had gone unfulfilled.[52]

Then, in 2009 the government finally appropriated funds for the permanent building, thanks to Department of Energy chief, Steven Chu, who is closely tied to a host of green researchers through his work at the helm of Lawrence Berkeley National Laboratory. It was a sign that solar energy—in all its forms—has scratched and clawed its way back up to and past the level of respectability it once enjoyed.

The National Renewable Energy Laboratory's permanent building is that "showpiece," even if it was built thirty years later than anticipated. It's the most energy-efficient commercial building in America and shows off all the technologies and modes of analysis that have been developed at NREL and other solar-friendly institutions.

NREL's analysts now routinely put out the types of pre-planning documents, forecasts, technology roadmaps, and promotional materials that characterized the golden age of commercial nuclear power in the late '60s and into the 1970s. Solar and efficiency entrepreneurs and researchers finally have the institutional support long enjoyed by nuclear power and fossil fuels in America.

When the National Renewable Energy Laboratory employees moved into the new building in 2010, it became official: After thirty long years, solar is no longer a renter in the federal government. It finally owns its house.

chapter 14

The Meaning of Luz

SOLAR-CONCENTRATING TECHNOLOGY has been anointed as perhaps the most promising large-scale renewable energy source in the world. Everyone from Google's chief climate officer to experts at the National Renewable Energy Laboratory to Southern California Edison agree that if the United States is going to generate a lot of electricity from the sun, solar thermal technologies are going to be the dominant technology in the near term.

In the green-tech boom of the first decade of the twenty-first century, billions of dollars flowed into companies that said they could make electricity from the sun's heat. By 2009 ten gigawatts, several large nuclear plants worth of capacity, were planned for the United States alone. The world's utilities were signing deals left and right while investors poured money into the companies they thought could deliver.[1]

Some projects were planning to use the "parabolic trough" variant of solar thermal technology. In this, long rows of mirrors curved like oil drums cut in half concentrate the sun's rays on a heat collecting tube, which is filled with synthetic oil used in chemical and nylon production.[2] The liquid flows to a heat exchanger, where it comes into contact with water. The water boils, becoming steam, which drives a turbine that generates electricity.

It sounds complicated, but the principle isn't very different from that used by little boys to torture ants with a magnifying glass: If you can capture and concentrate solar heat over a wide enough area, you can make anything get really hot. The Odeillo-Font-Romeau solar furnace in France, which was completed in 1970, is capable of generating temperatures of almost 7,000 degrees Fahrenheit. Trough power plants typically operate at far lower temperatures.[3]

The world knows the trough technology works thanks to the pioneering efforts of the only company to commercialize large-scale solar power in the twentieth century, Luz International. From the mid-1980s through 1990 the company built nine Solar Electric Generating Stations (SEGS) in the Mojave Desert, which have a combined peak capacity of 354 megawatts. The last two projects, each 80 megawatts, remain the largest solar plants in the United States. It's impossible to imagine a solar thermal industry without Luz's efforts.

A National Renewable Energy Lab analyst found that "The SEGS experience has been crucial and invaluable as a basis [for] what is happening today. Luz deserves great credit for its contribution to jumpstarting this technology."[4] Even today, the Department of Energy's Web page explaining the parabolic trough solar thermal technology notes, "Luz system collectors represent the standard by which all other collectors are compared."[5]

Beyond the hardware, the experience that engineers gained at the SEGS plants is the basis for the entire industry. One NREL report found that all the companies offering trough power plants in the market in 2005 had "ties through equipment and expertise to the SEGS plants."[6]

Three of the largest and most promising solar thermal players, Abengoa, Solel, and BrightSource Energy, are intimately tied to the Luz experience. Abengoa's vice president of technology development, Hank Price, was chief engineer at the SEGS.[7] Solel grew out of the ashes of Luz and drew directly from the technical knowledge base created at the original company before being purchased by Siemens for $418 million in October 2009.[8]

Then there is BrightSource, which is so much a part of the original vision that it was once known as Luz II, even if they have decided to use a different solar plant arrangement. Arnold Goldman, Luz's founder, is its chairman, and his first engineering hire, Israel Kroizer, serves as BrightSource's chief operating officer. Thirty former Luz employees now work for BrightSource.

Luz is the seed from which an entire industry has sprung. Whether concentrating solar power becomes "the technology that will save humanity," as energy policy analyst Joe Romm has anointed it, remains to be seen, but it's going to be a major part of any green tech–heavy future.[9] Yet the company's success is so improbable, so stupendously unlikely that it makes one wonder what options the world would be looking at if it hadn't been for Arnold Goldman's preposterous vision of a solar-powered world.

Arnold Goldman is exactly what one might expect an engineer-visionary to look like. He's small and a little round, with a beard that tracks his jaw line. He wears no moustache and he tends to talk quietly, except when he breaks out into peals of laughter. He laughed the hardest when I asked him how many people thought he was crazy when he told them what Luz was going to do. Everyone, he said.[10]

They might have been right. The audacity of Goldman's original plans boggle the mind in a green-tech world awash in Johnny-come-latelys looking for fast money in a growth industry. His company grew out of nothing less than a plan to construct a solar-powered utopia just outside Jerusalem. After selling his Los Angeles–based word processing company in the late '70s, he had moved to the ancient city with his wife and three children. He wanted some time off to clear his mind and finish a book that he'd been working on for more than fifteen years. The magnum opus was entitled *A Working Paper on Project Luz*, and it contains a philosophical-spiritual blueprint for all that came afterward.

In Hebrew, *luz* means many things. Most simply, it's the word for almond, but nothing is really simple about the word. It can denote the place in Canaan where Jacob had his vision of a ladder covered in angels climbing into heaven and back down again. "How awesome is this place! This is none other than the house of God, and this is the gate of heaven," the Book of Genesis records him saying.[11] The name Jacob gave the place, Beit El, means house of God. The connection to solar power, which uses energy arriving from space to make electricity on earth, is obvious. But there are more oblique and intriguing meanings.

In Jewish mystic texts, the meaning of Luz extends far beyond the story of Jacob's Ladder. Luz can also refer to a mystical place that is hidden from view, reachable only through a cave associated with a particular almond tree. Some say only true words can be spoken there.[12] One modern Rabbi has written that

This city was hidden from outsiders and even seasoned warriors could not penetrate its exterior. The singular passage to Luz was camouflaged by the foliage of an almond tree whose hollowed out trunk led into a cave that led into a corridor that led into a city of spiritual light. The City of Luz was impervious to destruction by conquering nations and its inhabitants continuously lived as a cohesive community throughout the millennia and were neither driven from their home nor ever assimilated or lost their unique purpose.[13]

Furthermore, the luz is a mythical bone in the body located somewhere in the lower back, from which some sects believe the Jewish people will be reincarnated.[14] Some rabbis argue that as the Jewish people will be rebuilt from the luz bone, the world will be rebuilt from the city of Luz.[15]

In most corporate histories, perhaps the company's name would not bear such deep inspection, but Arnold Goldman did not intend Luz to be taken lightly or superficially. His working paper is a two hundred–page tract that attempts to unite modern observations of space-time with the Torah's teaching to create a new ethical framework for life.[16] Along the way, he envisions a new Israeli solar-powered utopia.

He pleads for a new type of intellectualism he calls biocosmology, which will allow humans to "clearly see the relationship between the whole and the particulars of daily life."[17] Goldman's detailed description of Luz does not adhere to the conventions of utopian writing; it does not skimp on the practical. In Luz, there are companies and people who have regular jobs, but everything from product design to energy usage to corporate shareholder structure is informed by a devout belief in the unified (infra)structure for the universe. It is an engineer's paradise.

Goldman describes Luz as a set of two interlinked communities. Residents would live on a seven-year cycle, spending five years in the urban Luz, one year doing what they wanted, and one year at Luz North, an agricultural settlement. They would work four days a week at their regular jobs and two days a week developing "a concept which they believe increases their understanding and personal relationship with the unity and wholeness of the universe."[18] It would have been a bit like Google's 20 Percent projects, which allow employees to dedicate one-fifth of their time to self-chosen projects, but on spiritual steroids. Corporations would not exist in perpetuity, but would be sold after fifty years, with the proceeds disbursed to investors and employees on the basis of their service and value to the entity.

The urban settlement would be dense. One-third of a mile long and 1.2 miles wide, it "can easily be traversed on foot." No cars are needed for travel within the city. All automobiles are parked underground. For those who need to get around quickly, electric carts, "similar in design to some of the cars used on the more elaborate carnival rides," would provide transportation.[19]

In Luz, knowledge about the world would have to extend from the very natural—the sun and plants—to the human and technical. His goal was not to abandon the interactions and technology of the world so much as integrate lives that had been atomized. "It is an underlying supposition of Luz that the ideal of man should once again be the universal being who now enjoys and utilizes the fruits of the industrial revolution for his benefit," Goldman wrote. "The man which Luz foresees is master of his destiny and not a small inconsequential cog in the vast social-industrial machine."[20] To understand life, Goldman feels we need to learn it with our own hands at every level. So instead of getting our car repaired by a mechanic, the mechanic teaches us how to fix it ourselves. In Luz, a "garage is not viewed as a place to have one's automobile repaired; rather it is a place which teaches the owner how to repair the car, providing the specialized tools and parts needed to accomplish the task."[21]

This belief in fractal autonomy is where Goldman's interest in solar power arose. Whereas fossil-fuel markets are global, hopelessly beyond the control of hands-on understanding, Luz's energy systems are community run. Residents could learn how energy works, and in the small things they would be able to grasp the big things:

> Confronted with the question, 'how shall we heat our homes?', we must search for completely different solutions than conventional wisdom would dictate. We must ask a different question. It is no longer, 'how shall we heat our house?' but rather 'How can we heat our house and at the same time learn more about the unity of the world while maintaining control over our personal situation?[22]

There's little doubt that Goldman's ideas during this period were influenced by the *kibbutzim* movement at the time in Israel. During the 1970s thousands of volunteers flocked to small communities in Israel to participate in agricultural and artisan workshops. First set up as socialist communes in the early twentieth century, they had survived to see a social era that found them fascinating.[23] While the 1970s were christened the Me-Decade in the United States, the social world of the Israeli communes was different.[24] People living on *kibbutzim* were expected to put the community first in all things, but particularly their job performance. Work was expected to serve two purposes: first, the making of stuff, and second, the remaking of one's self. "The instrumental aspect of work, where labour is seen as a source of livelihood and means for achieving material resources, is distinguished from the expressive aspect that relates to work as a part of one's inner life, a moral obligation, a means for self-realization and regulation, and even a path for personal salvation," one kibbutz scholar explained.[25]

But Goldman is not and was not soft-headed. He knew that the main purpose of a heater is to make up for the heat a home loses. Gas heaters and heat pumps could keep a place warm, but "these solutions remove us as observer and participant in the basic energy flow process, confirming our dependence on a complex energy distribution process totally out of our control."[26] Though his justifications might be philosophical, his solutions—modular prefab construction, better insulation, solar planning in the siting of the house for southern exposure, solar hot water heating, and a greenhouse attached to the home—were practical.

Goldman argued for energy autonomy, but only at the community level—not for individuals. Unlike some of his American counterparts, he welcomed the reliability that came with networking power sources together. But centralized power generation had to fit into the Luz framework. To accomplish this, a wall "composed of a long series of parabolas which focus light on an energy-absorption tube" would enclose the cities.[27] Of course, no company existed to actually build such an energy conversion machine, so Goldman proposed the creation of International Solar Equipment Corp. The company "builds a sound product line from both a business and ethical perspective, designed to convert incident sunlight into a convenient-to-use form for home, commercial and industrial customers."[28] It was to form the cornerstone of the community, employing its residents and providing its power. As the 1980s opened, the company began to dominate Goldman's thinking. Luz didn't have to be a city. He himself had described it as "not so much a city as a thought or phenomenon." Luz the place was abandoned, but Luz grew. In the early '90s, the company made such a large percentage of the solar electricity in the world that it effectively was the entire solar industry.

"THE TECHNOLOGY THAT WILL SAVE HUMANITY"

It is not every day that a successful energy company grows out of a vision to unite the scientific observations of space and time with the Torah and human behavior. But that's how Luz started. Goldman did some of the initial business planning out of his garage and then hooked up with a French-Israeli financier, Patrick Francois. They set up a small office in San Francisco and started looking into how they were going to finance their projects.

At first, their plan was to provide heat, not electricity, for textile mills. They figured they could simply guarantee that the cost of their power would be 10 percent less than the natural gas power the companies already used.[29] After forming a U.S. corporation and Israeli subsidiary, they went looking for investors who might back the plan. It was an unusual company to say the least. If the investors asked solar experts about what the American-Israeli was planning, they would have found Luz's plans far-fetched: No one had ever built a solar thermal plant like the one they proposed.

Goldman, however, had earned some trust from the previous investors in his software company, so some of them came aboard. His pitch was also particularly attractive to American investors who wanted to help grow the Israeli economy. After all, the Middle Eastern country had plenty of sun but no oil. His argument didn't focus on the cost of the technology but instead the local benefits it could provide to Israel as a homegrown manufacturing concern.

It was a good time to be making such arguments. The energy crises of the 1970s reset the world's expectations of its energy supply, as the geopolitical shocks to the system exposed the world as dependent on fossil fuels. In the United States, millions reconsidered the lives they had built on assumptions of never-ending, always-on oil. The effects were even more acute in Israel. Solar based–energy self-sufficiency seemed like a great idea to big-picture investors and the Israeli government.[30]

With the only tangible evidence of his business a model that could fit in a suitcase, Goldman struggled to find traditional venture capital investment. Eventually, he decided to go to individuals for help. Many people interested in helping Israel hadn't thought to invest in the company's industries; instead, they supported the country through philanthropy. Luz, in its way, helped change that. We don't have a technology to show you today, he told potential investors, but give us some cash and come by our headquarters in six months. They'd show a prototype then. Led by Newton Becker, a philanthropist and prominent member of the Los Angeles Jewish community, they bought it.[31]

Then, all he needed to do was get a headquarters and build a prototype—in six months. Luckily, Francois, who been dispatched to Israel to find technical talent, had come across a brilliant young engineer named Israel Kroizer, who had studied heat transfer in solar systems at Technion, the Israeli Institute of Technology. Kroizer plunged into the work. He's now the chief operating officer of BrightSource, but at the time, he was just a kid who wore shorts and sandals and read stacks of books on energy systems. With a tiny team, they worked around the clock to build a prototype on the roof of their new office in the allotted time.

Luckily, they weren't building from nothing. Frank Shuman, a successful Philadelphia inventor, had shown that parabolic troughs could be used to focus the sun's rays and drive an engine. His 1911 pitch to investors could have been a model for Goldman's own. Like computers in the early '80s, flying machines were the unlikely-technology-made-real of Shuman's time. He wrote that

> you will at once admit that any businessman who was approached several years ago with a view to purchasing stock in a flying machine company would have feared for the sanity of the proposer. And yet, after it has been shown conclusively that it can be done, there is now no difficulty in securing all the money that is wanted, and rapid progress in aviation is from now on

assured. We will have to go through this same course with solar power and I am confident we will achieve the same success.[32]

Shuman's pitch worked and he eventually built the world's first solar power plant in 1912 in Meadi, Egypt. Its sixty-horsepower engine was used to pump six thousand gallons of water per minute to nearby cotton fields. World War I, however, put an end to his solar projects. Researchers like Farrington Daniels at the University of Wisconsin picked up solar research in the years after World War II.[33] In particular, the design of parabolic collectors that could track the sun and selective coatings for capturing more solar energy drew attention in those years.[34]

Work began in earnest in the early 1970s and intensified after the energy shock of 1973. Aden and Marjorie Meinel proposed using solar farms consisting of parabolic troughs focusing the sun's rays on a tube of energy-absorbing liquid, and multiple researchers explored the possibilities of different trough designs.[35] Perhaps Luz's closest ancestor was the Sandia National Laboratory's Solar Total Energy Program, which, like a highly technical, government-funded version of Project Luz, imagined the growth of one thousand home communities that made the "maximum possible use of the energy collected by a system of solar collectors."[36]

By 1981 a review by the Solar Energy Research Institute (SERI), later renamed the National Renewable Energy Laboratory, found that after just a decade of research into solar concentrating technologies, they had a chance of becoming as inexpensive as the gas and coal plants used to produce power during times of peak load.[37] But government solar programs were harshly criticized because most contracts went to big R&D and aerospace companies that, critics contend, never intended for solar energy to succeed.[38] They did, however, produce valuable baseline cost and performance estimates that reduced the uncertainty associated with the technologies, even if it bogged down the "solar revolution" in endless bureaucratic technical reports. The greater problem was that bringing a technology to market is expensive and there was actually very little monetary support for solar thermal R&D. In 1981 SERI spent $11 million on solar thermal energy systems out of the lab's total budget of $160 million. That same year, total civilian R&D expenditures by the government totaled $15.3 billion. The military spent an additional $17 billion. Given what government researchers and contractors had to work with, they provided a valuable service in generating information for the public domain on which entrepreneurs could build.

Public research, whatever its flaws, provided a map of the path of least resistance to commercializing solar power. The 1981 SERI review found that parabolic troughs were the most mature technology out there,[39] and after a crazed six months, Kroizer had built a working prototype of a parabolic trough system on the roof of Luz's Israel headquarters. The investors dropped by, re-upped, and Luz was on its way.

Their sales team fanned out across the American South, talking with textile factory owners and offering them energy at 10 percent less than they were paying, guaranteed, if they could spare about seven acres for the fields of mirrors. By the end of 1981 Luz had built two plants in Georgia and one in North Carolina with the help of investors enticed by the 25 percent tax credit for solar installations.[40]

That's when the company hit its first snag. The business model wasn't working. They had to talk with way too many companies to make sales. Sure, they had three customers, but they had to talk with a hundred companies. Their first plants were seriously underperforming, too. As Goldman put it, "The sun didn't work like we thought it would." Even haziness really negatively impacted the performance of the plants, and the southeast was hotter than it was sunny. Predicting how much energy a particular plant would produce was more complicated than it had seemed.[41]

The Luz dream could have died right then.

But as it turned out, the U.S. Congress had provided a perfect opportunity for a company like Luz—as well as a variety of less savory characters—through a piece of legislation called the Public Utilities Regulatory Policies Act, or PURPA, as it's known to policy wonks.[42]

In broad strokes, PURPA created a guaranteed market for any electricity that a so-called "qualifying facility" could produce at what was known as the utility's "avoidance cost." To qualify, you needed to generate electricity from a renewable source or to output both heat and electricity from the same power plant. There were various other stipulations, not all of which helped independent power producers. The original legislation limited the size of facilities to just thirty megawatts.

Luz's power plants could have been much more efficient at larger sizes, like other thermal power plants, but the size ceiling was eventually lifted to eighty megawatts and then discarded altogether in 1990 with some help from Luz lobbyists. At a hearing on lifting the size restriction, Senator Tim Wirth from Colorado noted "there was no policy rationale for limiting any of the small power producers."[43] The limit was arbitrary, or as Goldman saw it, "There was not a lot of interest in us getting too big."[44] On the whole, though, PURPA, along with the tax credits provided by the federal government, provided the market opening through which Luz squeezed in the 1980s.[45]

First, however, they had to prove to outsiders that they could make electricity. The company decided to contract with Bechtel to do a feasibility study that could be passed on to their prospective customers and investors. When the day came to present their technology, twenty Bechtel engineers filed in with coats and ties. Luz sent a single engineer, the young Israel Kroizer, wearing sandals and packing in his stack of books.

After a long meeting, Bechtel's engineers left impressed and helped complete the study of a design of a power plant for Southern California Edison in the Mojave Desert.[46] On Valentine's Day 1983, Luz signed a "sweetheart" contract with Southern California Edison for its first solar electric generating station. Though the cost of electricity that first plant produced was higher than avoidance cost, Southern California Edison rolled it up with a second plant so that the average of the two deals met the avoidance cost requirement.[47] Determining what exactly avoidance cost should be was complex, but it was supposed to reflect what it would cost the utility to provide the same power over thirty years based on standard cost projections.

Although PURPA was a piece of federal legislation, it left implementation up to the states, and California's Public Utilities Commission (CPUC) had made it clear that it expected utilities to do as much as they could to buy power from qualifying facilities. The CPUC told Southern California Edison and others that they had to offer a variety of formulaic contracts to the small power producers to offset the utilities' enormous bargaining power. In effect, the type of agreement Luz signed early on— known as Standard Offer Number Four—was a feed-in tariff. The solar company was guaranteed a fixed price for its electricity for a whole decade. Contracts in later years were not offered on such excellent terms, as the price the utility paid was pegged to natural gas prices, which plunged in the middle of the '80s. But for the first few years, there was an excellent opportunity to create a business out of renewable power, regardless of short-term fluctuations in the price of fuel.[48]

While the utility knew it had to make nice with would-be independent power producers, they were not always sanguine about the intrusion on their business. In a 1987 annual report, Southern California Edison was none too happy about being "forced" to buy "power, which, in most cases, was not needed and at costs much higher than the electric power the Company can now produce."[49] Luz's facilities produced power exactly when the utility needed it, albeit at a higher cost. On the other hand, they were also a hedge against what many assumed would be much higher energy prices.

While the government side of things was getting worked out, Goldman and Francois continued to get the projects financed with a combination of luck and skill. Their investors, who had become committed to the project, put up personal guarantees for construction of the first plant. They offered suppliers stakes in the project if it failed to be financed. Every cajole and incentive in the deal-making book was necessary to get the initial deals funded. The fact that Luz had Newton Becker, one of the wealthiest guys in Los Angeles, as chairman of the board, helped. It may have helped more that Luz became one of Israel's largest exporters of technology. The reasons to invest in the company extended beyond the merely financial, although those incentives were there, too.

From 1982 to 1991 Luz raised more than $1 billion dollars to build its plants. And the investors on those plants made money. The key to the deal, though, was that Luz had to put the plant into service by December 31 of a given year so that its investors could use the tax credits their work generated.[50]

The first plant had to be online December 31, 1984. It came online on the 27th, with only four days to spare. Although there may have been roars of happiness emanating from the control room that day, the yearly sprint to finish would eventually take its toll on the world's most prominent solar company.[51]

THE RIDICULOUS PROJECT

That first plant that Luz built produced fourteen megawatts. Its power was incredibly expensive, running close to 25 cents per kilowatt-hour, four times more than natural gas.[52] But it was the first of its kind and everyone expected the costs to run high. Just getting to that point had been incredibly difficult. Goldman's engineers referred to the plant as "The Ridiculous Project" because building a first-of-its-kind solar plant that would make electricity at 25 cents a kilowatt-hour was nearly impossible at the time.

Luz and its investors realized that SEGS I was going to be very expensive. The real questions would come in the second and third plants. Government reports predicted that costs would drop as more plants were made. Could Luz make its technology cheaper by using, learning, and developing it?

Ultimately, that's the bet for all renewable technologies and the reason why they are supported with tax credits across the world. If there is a market, the thinking goes, companies will compete to make their technology as cheap as possible while trying to grow larger. With most renewable power still priced far above its fossil competitors, many renewable technologies have to cut their costs in half or more.

So in 1983, when the second plant came online, it was tremendously exciting: It made electricity from the sun at 16 cents per kilowatt-hour. That was the kind of leap the company had to make. By the time reporters descended en masse on the Luz facility in 1989 in the wake of the Exxon Valdez oil spill, their new plants would be making power at 8 cents per kilowatt-hour.[53] The problems and bugs were being engineered out. Their scientists were finding more efficiencies. Technicians deployed new solar collectors. R&D developed new coatings for the solar collecting tubes along with a new, higher-temperature liquid for transporting solar heat through the plant's tubes. Luz was riding the cost curve down—and soon, they figured, the electricity they generated would be competing on equal footing with the natural gas and oil plants that powered California.[54]

And why not? Goldman asked himself. Why couldn't technology solve the problem of generating electricity without generating pollution?

During his time in the computer technology world, he had seen the power of the peculiar technological faith that underpinned Moore's Law, a famous dictum that the number of transistors and, hence, available computing power that could be put on a chip would double every eighteen months or so.

Though Kevin Kelly would likely disagree, there is no particular reason that Moore's Law be any kind of "law" at all. It's no more written in the universe than a man-law from a beer commercial. At first, it was just a tendency of the technology, but as it held, it became a core belief of the semiconductor companies and they organized to maintain it. Goldman said that

> there was no reason that Moore's Law should have been able to happen, but people believed it. If you were dealing with that as some bureaucrat, you'd never be able to take it seriously. But it was an outgrowth of people believing—the different kinds of machines, technologies, mathematics, massive stuff that feeds a vision that has modified the world in incredible ways.[55]

And Goldman believed. "I came from that environment and I bought into that learning curve."

It required, one might say, faith. "Arnold is a mystic," said Michael Lotker, the former vice president of Business Development of Luz turned Rabbi who wrote the defining report on Luz's corporate legacy.[56] "He really had the interest of the world at heart." Although Goldman read critics of technology like Lewis Mumford as well as techno-optimists like Herman Kahn, he preferred the philosophy of humanist Eric Fromm and author Joseph Chilton Pierce, who, he believed, showed that people could "blend the knowledge and experience gained in industrial revolutions to understand and control their lives," Goldman wrote. "It is the belief of this group of people that the knowledge and technologies gained thus far can be utilized to advance man to a new level of wholeness."[57] Luz was the expression of this peculiar kind of considered technological faith: Industrial infrastructure could better humanity, and technology could work for humans and not the other way around.

The company's string of triumphs, and access to investors, made Luz a very prominent member of the solar industry. It was basically the only large solar company. This was tough because it couldn't bank on an ecosystem of companies to purchase the components of plants like mirrors and solar receivers from their suppliers; instead, Luz had to support an entire value chain of companies all on its own. It was also responsible for lobbying on behalf of an entire type of energy that had few friends in high places during the Reaganite '80s.[58] During that time, energy prices were low as the malaise of the 1970s had been postponed.

So lobby it did. Luz representatives appeared before Congress a half dozen times during the company's run.[59] Two issues consumed most of their time. First, the solar investment tax credits were extended year by year. This meant that planning long-term was difficult for Luz. The company asked time and again: Why not provide long-term guidance? Second, the cap on the size of the facility they could build really began to hurt Luz. All thermal power plants work better at larger sizes. This scale law is why fossil fuel plants are rarely built any smaller than a few hundred megawatts these days. Luz had been limited to thirty-megawatt plants until 1987 and then eighty megawatt plants thereafter. But why have a limit at all? By 1989 Luz was convinced that if they could build a power plant of normal size, they could be cost competitive with natural gas as soon as the plant was built. In 1989 Joshua Bar-Lev, Luz's general counsel, argued before the Senate that

in five years we have brought our costs down. In five years we have shown that we can build on time, on schedule, and we are telling you that if you lift these limitations in a very short amount of time . . . we will be cost competitive. In other words, what I am saying is that if you look at it in terms of a decade, in one decade, we believe that we could start from zero, and produce for you a cost-competitive energy technology.[60]

But that was not to be. In 1989 Congress extended the solar tax credit for 1990 by just nine months instead of a year. The crazy deadline Luz had just been hitting had suddenly been moved up three months. In racing to finish the plant, they incurred $30 million of cost overruns after previous plants had come in on budget. At the same time, the Republican governor of California, George Deukmejian, vetoed a property tax break in the waning days of his administration that Luz had received up until that point. Nevermind that Luz's eighty-megawatt plant would pay four times the property tax of a comparably sized natural gas plant, Deukmejian wasn't about to give a tax break to a solar company.

The company was seriously hurting as they staggered into 1991, but they had some hope, at least in California. A Republican, Pete Wilson, had been elected governor and he knew Luz well, having helped secure the investment tax credits that made their projects financeable. So Luz kept building its tenth solar power plant, despite its precarious situation and the lack of the property tax exemption. "We had 2,500 people working in California. Wilson didn't see any reason the legislation shouldn't go through, so we decided to keep the project going," Goldman recalled. "He said he'd try to get it going on the fast line, which meant it'd be through by March. If we got it passed, we could get the project funded."

Luz would live another life.

A CRIME OF NEGLECT

In the spring of that year, with the fate of Luz in the balance, the bill came before the State Assembly and Senate. It had broad support, having passed the year before with 90 percent of the Assembly voting for it. The passage of the legislation seemed assured.

But while Luz's pleas felt righteous from the inside, outside observers were less kind about the tax break Luz was trying to hand itself. In 1991, as the property tax provision came before the California government, the state faced a projected budget shortfall of more than $14 billion.[61] Though the projected cost to the government was fairly small, something on the order of a few million dollars per year, Luz's lobbying methods drew attention from the reporters paid to watch the state capitol building.

At a hearing about the bill, Tom Bane, a Democrat with strong ties to Willie Brown, then speaker of the state house and a politician who did not have the most ethical reputation, showed up to support Luz. His plea, as recorded in the *Sacramento Bee*, was not exactly the sort of endorsement a company wants splashed in the metro news section.

"I know a lot of the people who have invested in this company," said Bane during the hearing. "If this project goes down, they're going to get nothing."[62]

The *Bee* further characterized the bill as "among the most heavily lobbied bills of the year so far, with squads of lobbyists and public relations experts scurrying around the Capitol on its behalf."[63] The Israeli government was lobbying on behalf of Luz. Willie Brown's folks were lobbying on behalf of Luz. The Sierra Club was lobbying on behalf of Luz.[64] Everybody, it seemed, thought the property tax break should pass. And that smelled fishy to *Sacramento Bee* journalists. Dan Walters, who is still with the Sacramento paper, recalled,

I think the attention was sparked mostly by the fact that it was a one-company tax break of serious proportions while the state was experiencing a budget crisis that was, proportionately, worse than the one it now faces and that it was getting an extraordinary push from liberal Democrats who ordinarily opposed such corporate tax breaks as they bemoaned cuts in health, social services and education spending.[65]

Solar or not, "soft" energy or not, Luz was playing hardball to get its property tax exemption. And no wonder: Over the course of the project's thirty years, paying the state the money would have added $90 million to its costs, the company's chairman estimated.[66] In the scheme of the California budget, the cost was small, but in the scheme of Luz's business model, it was make or break.

That was Luz's claim at least, one that was sadly borne out by history. But the state's finance department didn't see it that way. "We're not convinced the plant won't be built anyway," a California Department of Finance aide testified before a state Senate committee. She also knocked Luz's other argument: that the state would lose the payroll and sales tax revenue from the project if it wasn't built. She said the investment money would just be spent somewhere else in the state.[67]

Democrat Phil Isenberg, representing an area of Sacramento, hit the group the hardest, calling attention to the giveaway and terming it a "moderate-terrible" bill.[68] He echoed Phil Stark, who blasted Luz in a 1989 *New York Times* article about federal tax incentives. "To the argument that Luz produces clean, nonpolluting energy, Stark responds: 'So do squirrels running a treadmill. It's a question of how much you want to subsidize the squirrel-breeding industry,'" he said.[69] Thus, solar electricity, as produced by Luz, was not seen as needed by the country. It was squirrel-breeding, not heavy industry. And yet it was heavy industry. During the construction of the final Luz plant, it was one of the largest building projects in the state.

Despite all the fuss, the two chambers of the California legislature passed the tax break with only eight dissenters between them.[70] The bill went to Governor Pete Wilson's desk. Goldman says that Wilson had told the company he would sign the bill.

Then came the big story on the passage, which ran under the headline, "Tax Break Quickly OK'd For Solar Energy Plant—Bill Assailed as 'Giveaway' Goes to Wilson." The story recorded how Democrats split, fighting amongst themselves. Two days later the *Sacramento Bee* ran an editorial under the headline, "Luz Casts a Long Shadow."[71]

The *Bee* pulled no punches:

> Luz isn't operating so close to the edge of financial extinction that it hasn't been able to afford to spend tens of thousands of dollars on political contributions in recent years, hire some of the highest priced lobbyists and public relations consultants around the Capitol, and pay for a fancy party for the Energy Commission staff last year, not to mention the $1 million it's paying San Bernardino to keep silent.
>
> Sacramento Assemblyman Phil Isenberg calls SB 103 a moderate-terrible bill, not a major one. But at a time when the state is facing a $13 billion deficit, that's bad enough to merit a veto.[72]

With that on his mind, Wilson decided he couldn't sign the bill right away, despite what he'd told Luz.

"The exemption Luz wants isn't going to help the solar industry as a whole," the *Bee* editorial board charged. "According to the legislative committee analyses, Luz is the primary beneficiary." But that's because there was no utility-scale solar industry. Luz was the only company that had thrived during the Reagan years, and then they ended up penalized for their unique success. Wilson, new to the governor's mansion, asked that the bill be withdrawn, altered, and resubmitted. It was, the legislature passed it again, and Wilson finally signed it. Only five weeks had passed, but by then, the company had lost the participation of new investors. They ran out of money two months later.[73] The company entered Chapter 7 bankruptcy.[74]

After the company's untimely demise, Newton Becker, the company's board chairman, went before Congress to deliver a fiery oration about the irrationality of American energy policy. "As an investor I am here much like a crime victim trying to change a law," he said. His testimony, which quickly summarized the whole lot of unfairness and regulatory silliness, demonstrated the tremendous carelessness of U.S. energy policy. He noted that the time to plan for the energy supply of the future is when those resources are adequate and cheap. That also happens to be the time when Americans are most likely to ignore the problem. "Without greater attention by Government, we are doomed to breathing foul air and increasingly relying on unstable countries in the Middle East as a major source of our country's energy," Becker concluded. "There is no solar thermal electric industry today."[75]

The game was over for Luz but not for the Solar Electric Generating Stations. They proved indestructible. For a quarter of a century, they have produced power reliably. Learning in the field has brought operating expenses down. The plants make money. Many solar thermal engineers have trained out in the fields of the Mojave.[76]

With his company busted, Arnold Goldman hunkered down in Jerusalem. The tremendous success of SEGS was one of those stories that floated around the renewable energy world, as apocryphal as a real project can be. Luz claimed that their next generation of larger plants was going to be cost competitive with natural gas generation. Looking back at the beginning of the green-tech boom of the early twenty-first century, it's impossible not to wonder, "How did everyone let that happen? And what would have happened if it hadn't?"

After Luz, solar thermal deployment nearly ground to a halt for more than a decade. The cheap energy of the 1990s kept interest in the technology low. Then, in 2004 Goldman started to get excited again.

Oil prices began to climb again, and this time it looked to be more because of the tremendous demand coming online in Asia, not because of geopolitical problems in the Middle East. It seemed like the rise in energy prices could be permanent, or at least that was a pretty good thesis for investors to work from. At the same time, some form of global climate agreement seemed like a pretty good bet to drive the desire for low-carbon energy sources.

Goldman started to contact the old Luz gang, asking if they wanted to give the solar business a go again. The answer was almost always yes. Luz II was born. In 2006 Luz II changed its name to BrightSource Energy and picked up venture backing of $160 million over three rounds of financing between October 2006 and May 2008. (For the full Bright-Source story, see the very last chapter in this book.) In 2009 the company signed the two huge solar deals in the world for a combined 2,610 megawatts of capacity. One of them is with Southern California Edison, the utility that donated the land on which the very first Luz plant was built.[77]

It's as if Goldman planned it all. Luz has acquired the other definitions ascribed to the Hebrew word by the mystics: indestructible. It is the seed of a solar resurrection.

chapter 15

How to Burn a Biological Library

THE PLAYA LAKES west of the desert near where New Mexico, Utah, Colorado, and Arizona come together are some of the strangest biological environments in America. Small depressions in the prairies, with areas usually less than thirty acres, the lakes pop into and out of existence depending on the rain.[1] Year after year, the playa lakes cycle from dry to wet and back again. Inside these lakes, there reside vast populations of microorganisms that through the inexorable logic of natural selection have somehow adapted to an environment in which water, the key ingredient for all life, comes and goes with the capriciousness of a wandering thunderstorm.

In the early 1980s these microorganisms were what sent Jeff Johansen careening all over the ranches and farmland of parts of New Mexico, Colorado, Texas, and Oklahoma in his trusty Renault, armed only with some intuition about where these ephemeral lakes might be and a game willingness to hop barefoot into the water to secure samples of the organisms unique to that little puddle on the prairie.[2]

Every time he came across one of the round, briny lakes, he could be looking at a hidden black gold mine. It was possible that any lake could hold one of the most valuable organisms that a man could find. Inside some invisibly small creature's genes, there might be a recipe for how to synthesize something like crude oil in large amounts. With the right combination of nutrients acting like a genomic key, they might unleash an organism that reproduced quickly, squirting out fuel as it went.

Finding and growing the best algal strains for producing fuel out of carbon dioxide and sunlight was the essence of the National Renewable

Energy Laboratory's Aquatic Species Program. For nearly twenty years, in search of the few strains out of the millions on earth that might work the way humans wanted them to, the lab built a living, respiring library of carefully collected organisms. Phycologists like Johansen who study algae fanned out across the country looking in thousands of highway ditches, Southern estuaries, Hawaiian rivers, and everywhere in between. The best performers, though, came out of the tough neighborhoods of the playa lakes, where being able to produce fat may have helped save them from dying of desiccation. One of the very best species at converting sunlight and carbon dioxide into fatty fuel came out of a lake twelve feet across.[3]

The program was a bioprospecting effort like no other in our nation's history. Despite meager funding, the Aquatic Species Program, initiated under President Jimmy Carter, laid the scientific foundation for making diesel-like fuel from the fat that microscopic algae accumulate in their cells.[4]

Three thousand species were eventually isolated and cultivated. Now, nearly all of them are gone. Less than half of the fifty-one varieties that were carefully characterized as potential high-value strains remain in labs for study.[5] "Just when they started to succeed is when the plug got pulled," said Johansen, now a professor at John Carroll University. "We were growing them in ponds and we were going to grow enough to have them made into a diesel fuel."[6]

The program was part of the large investment that the Carter administration made into alternative energy in the late 1970s. Between 1977 and 1981 energy research and development accounted for more than 10 percent of the federal government's total R&D budget. Almost 25 percent of all the money the government spent on energy research between 1961 and 2008 was doled out between 1974 and 1980. In the overall context of the federal budget, defense spending, or even other government R&D, these investments, which peaked at $7.5 billion in 1980, are a pittance. But they provided energy researchers of all kinds a tremendous boost. The very idea that the United States should have a diverse energy portfolio was born during the era and provided a framework for investing in alternative energy sources outside nuclear energy and fossil fuels. Although it is certain that there was some waste involved in the Department of Energy's duct-tape and bailing-wire funding mechanisms, the R&D push made all kinds of progress on the energy problems that remain with us today.[7]

But when Ronald Reagan came into power in 1981, energy R&D funds dried up like a playa lake, leaving experts to find other work. As a result, knowledge seeped out of the nation's energy innovation ecosystem. The serious whiplash caused by the Carter-Reagan transition highlights the problems created by inconsistent funding for energy research. Even when money does flow into the coffers of energy researchers, those researchers worry. Most recently, President Obama has trumpeted the American Recovery and Reinvestment Act, also known as the stimulus package, as the largest increase in scientific research funding in history.[8] Scientists roundly applauded the billions of dollars that went into energy research, development, and deployment. But what about when the stimulus money runs out in 2011? "We're going to have to see what happens after these next two years, because what we need is not a drop, but a further increase in R&D commensurate with the task at hand," said Ernie Moniz, head of MIT's Energy Initiative at a Google-hosted panel on energy in late 2009.[9] The task at hand is transforming the American energy system, and Moniz knows that's exactly what did not happen in the last big energy R&D push.

Turning pond scum into oil isn't easy, but as a hypothetical energy system, it's elegant. The theory is that algae will produce more burnable fuel on less land than regular crops, perhaps something like a thousand gallons of oil per acre instead of a few dozen from conventional plants.

The food-versus-fuel debates that plague biofuels like corn-based ethanol would disappear. As about a third of U.S. corn production is now transformed into ethanol for blending with traditional gasoline,[10] that's a large enough percentage that critics have questioned the use of a food crop for energy production. One UN official called turning food crops into fuel "a crime against humanity" in 2007. Others have attacked corn-based ethanol as requiring too much fossil-fuel energy to actually make any sense as a replacement for transportation fuels.[11]

Conversely, algae could be cultivated on nonarable lands with briny water unsuitable for agriculture, using the carbon dioxide emitted from power plants as a food source. Oil produced with algae would neatly sidestep the problems that ethanol and other biofuels have run into.[12]

Some energy researchers had long supported the development of mass algae culturing, particularly of the species chlorella pyrenoidosa.

Vannevar Bush, Franklin Delano Roosevelt's science adviser and a key character in another chapter of this book, lent his full-throated support to the chlorella research from his perch atop the Carnegie Institution in the years following World War II. The organization, particularly scientist Herman Spoehr, played a key role in funding researchers and promoting algae as a human food, energy, or medicine source.[13] One study fed "plankton soup" flavored with a little salt to lepers in Venezuela who, the researchers reported in 1953, "drank it willingly. The taste, which was not unpleasant, varied with the species complex."[14] The Arthur D. Little Company's Flavor Panel, a group of American food testers, had a less pleasant experience. "The thawed, undried material alone showed vegetable-like flavor and aroma and was rated as 'foodlike and food satisfying' but with some of the 'notes' unpleasantly strong," wrote one of Little's consultants. "There was, however, a noticeable tightening at the back of the throat ('gag factor') and a lingering, mildly unpleasant aftertaste."[15] Further studies considered using algae as food, but perhaps with a little less verve than before. Eating algae became one of those things people might have to do in the sci-fi future, when the population of earth is so large that "a new source of foods or feeds is mandatory regardless of cost."[16] Still, the Carnegie-supported work had laid out some of the basic groundwork of what a system might look like for growing algae in large quantities.

Meanwhile, phycologists discovered that when they restrict algae's nutrients, some start cranking out far more oil than normal. Unfortunately, they also stop growing.[17] If the scientists could keep the algae multiplying and pull the "lipid trigger" anyway, they'd be in fat city. But their understanding of the biology was incomplete, and the task wasn't easy. Algal genomics were but a twinkle in the scientific eye. It would take some time and effort to know if and when the process would become cheap enough to compete with crude.

Another challenge was getting the algae to keep growing without injecting a lot of energy into the system. The government scientists installed large open ponds near Roswell, New Mexico, and began trying to produce tiny algae at oil tanker scales. It worked, but there were engineering problems. In retrospect, Roswell was less than ideal; it was just too cold.[18] Even under ideal conditions, though, the operation would have needed time to get all the bugs ironed out. Then the researchers could have begun to assess the real costs of growing algae.

But the program did not get that time (or money). By the time Bill Clinton took office, its funding had dwindled to a trickle, and in 1996 the Department of Energy abandoned algae altogether to focus all its biofuel efforts on ethanol.[19] A dark decade fell upon the field of algal biofuel. There wasn't even money available to take care of the algal collection that had been so painstakingly created. In an effort to salvage some of the science, a few hundred strains of algae were sent to the University of Hawaii, and a grant from the National Science Foundation was secured in 1998 to preserve the algae. However, when the money ran out in 2004, continuing the laborious work of maintaining the collection became difficult.[20]

And it is labor. The collection is really, truly alive and requires care like a tiny, slimy pet. The microorganisms sit in rows of test tubes, living and reproducing. Every two months they have to be transferred, or "passaged," to a new nutrient-rich tube. Random genetic mutations can enter a population and lead to permanent genetic changes. The algae can die.[21]

It's not exactly clear how it happened, but a review released earlier this year found that more than half the genetic legacy of the program had been lost. Only twenty-three of the fifty-one strains that were extensively studied during the program remain viable to grow again. The losses to the rest of the algal cultures in the collection were even worse. "The really bloody shame is that of those 3,000 [original species], there are maybe 100 to 150 strains that remain at the University of Hawaii," lamented Al Darzins, who heads up the resurgent algal biofuels research program at the National Renewable Energy Laboratory.[22]

The way the United States has spent green-tech R&D funding has hurt the efficiency of the research. We've gotten less for our money than we should have. Programs that started during the boom of the late '70s were abandoned during the early '80s, even if they were promising. Despite the best efforts of cash-strapped researchers, not everything can be frozen cryogenically awaiting the next period of high energy prices. If anything, government support should be highest during periods of low-energy prices. "We need a countercyclical investment pattern," said

Gregory Nemet, an energy innovation researcher at the University of Wisconsin-Madison. "When the market dies, we need to ramp up the public R&D. We need to grab those people and get all their data and codify their knowledge so it's not lost."

While the valuable NREL archive of algal biodiversity languished in the Hawaii basement, the world around it changed. Genetic and genomic research and understanding skyrocketed. Oil demand grew, particularly in massive developing countries like China, India, and Indonesia. We kept burning two barrels of oil for every one that we found.[23] All of a sudden, interest in algae-based biofuels exploded. Venture capital and corporate money flowed back into the field. On January 2, 2008, oil hit $100 a barrel for the first time. Despite some ups and downs, the price of oil remains substantially higher than it was through much of the 1990s. As a result, more than fifty companies are now at work on some aspect of biofuel production from algae.[24]

This time around, however, the big boys have even gotten involved. Exxon Mobil decided to invest $600 million into a joint venture with Craig Venter's Synthetic Genomics for research into next-generation algal fuels.[25]

Over the past few years Darzins has revived the venerable Aquatic Species Program at NREL. Applying new genetic science, they have been hard at work learning the biology of microalgae. Graduate student Lee Elliott of the Colorado School of Mines has collected five hundred new species as part of the long-term rebuilding of the biological library. To a certain extent, the problems of maintaining a microorganismal library have been solved. Cryogenic freezing techniques were developed at the University of Texas UTEX Culture Collection of Algae.[26] Because of this, the NREL team has been able to freeze and then revive 91 percent of their microorganisms.[27]

Despite the lost decade, algal oil makers are optimistic that they are about to ride a steep cost curve down to much, much cheaper biofuel. As they apply new biological knowledge and optimize growing algae, the cost will drop. As they capture economies of scale, the costs will drop again. In the best-case scenario, when all is said and done, algal biofuel could become a substitute for petroleum. But that won't happen anytime soon, and it could take a decade or longer.

Loving critics of the field like John Benemann, who worked on the Aquatic Species Program, say there are some legitimate reasons to be skeptical of algal biofuel's potential for large-scale oil production. The main problem is that nobody has been able to make fuel from algae for a cost anywhere close to cheap, let alone competitive with current oil stocks.[28] Some researchers question whether any kind of energy-conversion process based on photosynthesis will ever play a major role in our transportation energy system. Though most researchers disagree, one life-cycle analysis found that we would have to put more energy into algal biofuels than we could get out, a notion that, as noted, has also dogged ethanol.[29] Despite all the new companies in the space, none could be said to have succeeded. The prominent startup Green-Fuel, which grew out of Harvard and MIT research, went bust in May 2009 after blowing through $70 million.[30]

The point is that we don't know how well algal biofuel production might work when all the science and engineering work gets done. Although eighteen years of small amounts of funding yielded a lot of knowledge, it did not create anything resembling a commercial product or process. "The cultivation of microalgae for production of biofuels generally, and algal oils specifically, is not a near-term commercial prospect," Benemann maintained. "Larger-scale algal biofuels production still requires considerable, long-term R&D."[31]

Only $25 million was invested over the life of the Aquatic Species Program, which is just 5.5 percent of the total money the Department of Energy dedicated to biofuels over that time.[32] Adjusted for inflation, the program's total budget in today's dollars was less than $100 million. To put this tiny number in oil industry context, Exxon Mobil made $142 million in profit each day of 2008. Johansen, the phycologist who collected algae for the program, said,

They came up with this idea and in four years, they almost demonstrated the technological feasibility, and then the funding fell out. The maximum of funding was about $4 million a year. When I left, it was $800,000 a year. Now, there is all this biofuel work going on, and they are all going back to that public domain research. It kind of drives me crazy.

The neglect of the Aquatic Species Program and subsequent resurgence of algal biofuel interest is one of many examples that show how the lack of coherent, consistent energy policy has left the world's most oil-dependent nation scrambling in times of crisis. Johansen even went so far as to say that "if the Reagan and Bush administrations had not ended" the growth of the algal biofuels program, the United States would already have commercial algal biofuels.

Even under far less optimistic scenarios, if the Aquatic Species Program had been fully funded from its start until now, there is no question that we would know a lot more about the potential—and limitations—of algal biofuels. The point is that we do R&D as much to find what *doesn't* work as to discover what does.

Instead, however, we are left with some lessons learned, a partially missing library of microorganisms, and a lot of questions that investors and entrepreneurs want answered before the next oil price spike, whenever that may be.

IV.

Lessons from the Great Energy Rethink

chapter 16

What Happens When an Energy System Breaks

THE 1970S WERE A FULCRUM on which American society turned from one vision of the future to another. The long postwar boom gave way to energy shocks, polluted cities, and an unpopular war. The economy was stagnant; trust in government collapsed. The great protest movements and social achievements of the 1960s did not bring equality and justice for all. The idea of limits crept into the American mind, as unwelcome as a bad odor. As Vietnam exposed the limits of military power, other geopolitical events revealed the country's energy system had limits, too.

In October of 1973 the Organization of Petroleum Exporting Countries (OPEC), the cartel of oil-rich nations, asserted its global economic might by cutting off oil supplies to the United States. Although American support of Israel in the Yom Kippur War served as the immediate spark for the embargo, the crisis simply revealed the realities of the brittle American energy system.

Though few believed it, American domestic crude oil production had peaked in 1970 and was headed for a long, slow decline, despite the onrush of production from Alaskan oil fields. In the three years leading up to 1973, oil imports doubled to over six million barrels a day. A country that had gotten rich selling oil to other countries began importing vast quantities of the stuff from state-owned oil enterprises that were not afraid to use their mineral resources to achieve political ends.[1]

The reaction in the United States was swift. People were horrified. For the first time since World War II, Americans could not have as much fuel as they wanted when they wanted it. Rationing ensued. Frustration abounded. America could not control its own destiny. The civic and productive systems that we had built when oil cost less than water were far-flung, low-density, and energy-intensive. Muscle cars cost just a tiny bit more to drive than VW Bugs. How had America built its castle on a resource that seemed more like sand than rock? What had gone wrong?[2]

At the same time, a new movement was spreading across America. Spurred by the increasing recognition that human beings were destroying ecosystems and their own suburban backyards, modern environmentalism rounded into rough shape. "The residents of post-war suburbs lived in the midst of one of the most profound environmental transformations in the nation's history," historian Adam Rome wrote. "Every year, a territory roughly the size of Rhode Island was bulldozed for urban development."[3] The environmental problems caused by the sprawl became suburbanites' local touchpoints for the abstract notions of ecosystems and carrying capacities.

Tract homes were built as cheaply as possible, often beyond the reach of city services. Instead of paying for extensions of sewer systems, developers put in poorly designed and executed septic systems with predictably poor results. "In the late 1950s, for example, suds began to pour out of the faucets of thousands of suburban homes—the residue of nonbiodegradable detergents in septic tanks had contaminated drinking wells—and the resulting outcry helped make water pollution a national concern," Rome argued. In short, "The desire to preserve wilderness was the tip of an iceberg, the most visible part of a much larger concern about the destructive sprawl of urban civilization."[4]

Ecology, the science of studying networks of organisms, was exploding in popularity. As a result, many ecologists became prominent political figures. The conclusions of the time were unequivocal: "Do not disturb the balance of nature."[5] In the late '60s this information began to filter into the mainstream consciousness. In 1968 the *New York Times* ran an editorial by a local environmentalist under the headline "An Ecosystem Is a Partnership." The author wrote that

unfortunately, unless we heed the warnings of ecologists, and stop upsetting the precious balance of nature upon which this planet's web of life depends, all of us may be forced to live under giant astrodomes with the same kind of environmental controls developed for space travel. The meaning of an ecosystem will then be clear to everyone.[6]

The first environmental television series, *Our Vanishing Wilderness*, ran on National Education Television in October of 1970, a few months after Earth Day brought twenty million people together to celebrate a new environmental consciousness. The show brought the destruction of the environment home to suburbanites in much the same way that graphic footage added drama to the Vietnam War. The *Our Vanishing Wilderness* episode about the 1969 Santa Barbara oil spill, "the spark that brought the environmental issue to the nation's attention," was titled "Everybody's Mistake."[7] And the ecological position that "everything is connected to everything else," meant that vanishing wilderness might lead to a vanishing humanity.[8]

In short, the relationships between ecosystems, economies, energy, and technology became unstable. Sitting in the famed gasoline lines waiting to fill up their tanks, Americans of all stripes were forced to think more about where their fuel came from than they had in half a century. Many came up with some tremendously weird, inventive, and productive ideas. Most of the "new" policies and technologies bouncing around America in the first decade of the twenty-first century are milquetoast variants of the radical reexaminations of the country that occurred during the 1970s.

Yet in popular memory this great outpouring of intellectual energy has been caricatured and reduced to cartoon hippies playing with old windmills. However, there was much, much more to the era. Everyone had a plan for fixing the way the country used heat, light, and power. Like the raucous panoply of 1840s reformers responding to industrialization, the American energy system got its own constellation of would-be helpers, as the seeds planted back then bore decidedly strange fruit 130 years later.

Government and university energy specialists issued report after report, many pushing for large-scale projects like nuclear reactors that yielded more radioactive fuel than you put into them ("breeders") and synthetic fuel plants that converted coal into knock-off crude. Foundations convened large groups to discuss American energy: With a couple of exceptions, their sturdy, dour reports are like the existential opposite of coffee-table books. Formerly obscure researchers held conferences on "Energy Transitions" and argued about odd things like the return of feudalism as an unintended consequence of increased solar energy usage.[9] Thermodynamics, a musty old nineteenth-century science, was suddenly on the lips of physicists like Art Rosenfeld, who had formerly been more interested in quarks and gluons.[10] Ecologist Howard Odum put out books seriously suggesting apportioning UN votes "in proportion to the energy budget" of a country, including its haul from the sun and fuels.[11] Barry Commoner, an eco-socialist, published popular books attacking capitalism as the root of the environmental, energy, and economic crises racking the country. These were widely acclaimed by the mainstream media. The environmental movement figured out how to use bureaucratic levers to delay and block the building of power plants. Rodale Press, an early imprint for information on organic food, began to branch out into alternative energy. Magazines like *Not Man Apart*, *CoEvolution Quarterly*, *Clear Creek*, and *Mother Jones* came into being and emphasized deep environmental and energy issues.

While in a 2006 *New York Times* article, then-president of the Solar Energy Industries Association Rhone Resch declared, "We're not just hippies in garages in Berkeley anymore," the truth is that the people interested in alternative energy have never been "just hippies in garages in Berkeley." The roots of green technology spread wider and deeper. The choices society was mulling over during the 1970s were more nuanced than commonly assumed or misremembered. The technological ideas available were not simply buckydomes and solar hot water heaters on one side versus huge nuclear and dirty coal power plants on the other. There were branches and gradients running from tripped out solar home owners through the Amory Lovins's soft path and energy efficiency all the way to the grand solar farms of Aden and Marjorie Meinel.

There were so many interesting things going on during the 1970s that it would be impossible to cover them all even within the space of this entire book, so we'll look at just four key themes of green-tech practice and research that emerged from the period. Responding to the chaos of the times, different groups of researchers and tinkerers brought their own approaches to the energy problem. How they succeeded and failed hold some of the most directly applicable lessons for the current resurgence of green interest.

Following the line at the time that environmentalism needed to address the four P's—pollution, pesticides, population, and proliferation—we'll look at the four T's of '70s green: thermodynamics, transcendence, tools, and technology.

chapter 17

Thermodynamics

Thermodynamics is the science of energy conversion—how heat dissipates and work gets done. It's an odd science, rich with nineteenth-century language and personalities. Thermodynamics is, essentially, the study of how and why we can't build perpetual motion machines. Thermodynamics is the field in which entropy, the tendency that systems have to become more chaotic, was first recognized. Thermodynamics defines the boundaries for humans' ability to transform the real world.

But if thermodynamics can define the upper limits for the efficiency of our power plants and cars, it can also help us find the lower limits for energy consumption. And in a time when it seemed the nation's energy industry was out of ideas and money, a staid group of physicists used their knowledge to radically change the energy debate forever.

WHERE'S MY WEATHER CONTROL AGENCY?

In *1999: Our Hopeful Future*, John and Emily Future wake up in a wondrous world where John takes a helicopter to a thirty-hour-a-week job while the Regional Weathermaking Service generates snow for the pinochle game that Emily, a stay-at-home mother, is hosting. The book narrates, "Emily had the ladies out in the garden bubble—the new enclosed part of the yard (with dining terrace and thirty-foot swimming pool) where the climate was kept the same the year round. And the snow was beautiful through the clear plastic bubble."[1]

Timothy, their son, eats Super-Mishmosh cereal, which keeps him from ever having "a sniffle or cold." Nuclear and solar thermal power plants pump their world full of energy, and atomic aircraft cruise the skies. People live to be 115 and balding is a thing of the past. Finally, in January of that hopeful year, humans land on the moon.

The book presents a future that uses almost unlimited amounts of energy to remake the entire world for humans. Published in 1956, the author was no less than the prize-winning Victor Cohn, proclaimed upon his death in 2000 to be the dean of medical science writers.[2] The book grew out of Cohn's reporting on the scientific luminaries of the day, including Vannevar Bush, John Von Neumann, Buckminster Fuller, Linus Pauling, and innumerable other scientists at Cal-Tech and MIT. Cohn's book, which mixed fiction with reported profiles of possible new technologies, was like a distillation of scientific and corporate hopes at mid-century.

Never-ending scientific advance begets never-ending resources, which begets never-ending growth of the consumer economy of the post-War period so that everyone can share the wealth.

It's a vision, not to put too fine a point on it, of the American dream.

The country's leaders, however, created more serious and quantitative descriptions of this future. At a conference on "Energy, Economic Growth, and the Environment," hosted by Resources for the Future in Washington, DC, in April of 1971, Philip Sporn, the brilliant bespectacled chief of American Electric Power, presented himself to the crowd as a realist.[3] He was, after all, a man of science, an engineer, and a master builder of the postwar economy. His company was the one that had pushed the limits of power plant technology, making it bigger and more efficient. In the process, they drove down the cost of a kilowatt-hour of electricity.[4]

The words that came out of his mouth sounded scientific and grounded in the nitty-gritty work of keeping the lights on. He eschewed an estimate from the conference's organizers that the U.S. economy would use 190 quadrillion BTU of energy in the year 2000, predicting total American energy demand of 152 quadrillion BTU instead. At the time the United States was burning 68 quadrillion BTU, or 68 quads, of fuel. Sporn made clear that he was offering a sensible middle path. He wasn't a business-grubbing energy executive hoping that energy sales would be 190 quads. However, he saw real value in energy and didn't think it wise to limit energy usage, as advocated by some scientists like University of Massachusetts Amherst physicist, David Inglis. Sporn warned,

> I doubt that it is in the best interests of the American people to adopt Professor Inglis's program for limiting energy growth. Since every projection of population, GNP, and industrial production indicates major increases in total energy use, electric energy use, and per capita use, I shall discuss the impacts of environmental costs on the program of expansion and growth as I visualize that growth.[5]

In plain terms, Sporn dismissed Inglis. The weight of "every projection" added up to an inevitable future in which Inglis's wacky ideas weren't even worth discussing. In fact, Sporn had bigger fish to fry. Given that energy demand growth was inevitably going to reach higher—but not too high—he devoted the rest of his speech to dealing with the real business of figuring out, or inflating, the environmental costs of producing all that power.[6]

Who could blame Sporn? Every projection *did* call for major increases because between 1900 and 1970 average annual energy growth was over 3 percent. Electricity demand growth from 1950 through 1970 had averaged 7 to 10 percent per year.[7] Consequently, growth in energy usage was *a sure thing*. If growth like that was a sure thing, the only solution to the "energy problem" was to build ever more and larger power plants. But what if, just maybe, the forecasts Sporn and everyone else created were less accurate than they seemed? What if all those models produced by government officials, corporations, analysts, and academics were *all* wrong? The country's engineering elite would have spent a decade planning for the wrong future.

There were people like Inglis who believed that a different future was possible. Some very specifically called attention to the inadequacy of the top-down forecasting methods. For instance, Herman Daly, a University of Maryland economist, lobbied intellectual grenades at the energy industry, but to no avail. Daly wrote in 1976 that

> recent growth rates of population and per capita energy use are projected up to some arbitrary, round-numbered date.

Whatever technologies are required to produce the projected amount are automatically accepted, along with their social implications, and no thought is given to how long the system can last once the projected levels are attained. Trend is, in effect, elevated to destiny, and history either stops or starts afresh on the bi-millennial year, or the year 2050 or whatever. This approach is unworthy of any organism with a central nervous system, much less a cerebral cortex.[8]

Few in the traditional energy forecasting business listened. As it turned out, they were terribly, almost unbelievably wrong. In 2000 the United States used less than one hundred quads of energy.[9] In 2009 the United States used just 95 quads of energy. That's less than even the outlier/activist Inglis had thought possible. Yet no calamity befell the United States. In fact, the 1980s and 1990s are not exactly considered years of hairshirted deprivation nor rationing. We are more likely to remember them as the adolescence of the sports utility vehicle and the exurbs. They were a time when energy usage, supposedly, was profligate. Those on the left may think Americans live in a sprawled out, McMansion, and SUV-loving world, but that's nothing compared to what those "in the know" were projecting for the country.

Consider the high-end industry estimates, which would have required nearly doubling the current U.S. energy infrastructure. For every power plant, add another. For every ten cars, add nine. For every inflated exurban tract development, add another. It's hard to imagine how the world could have ever used all that energy.

As different as corporate projections and *Our Hopeful Future* might seem, they both failed because they rested on mountains of assumptions about what kinds of change were and weren't possible. The ways that Cohn's book deals with the future illuminates the innards of the corporate projections. Both types of documents imagined few disruptive changes. Commuters might trade in cars for helicopters, but they would still go to managerial jobs in suits, even if those suits were made of yellow polyester. Although the world of the Futures might seem radical, Emily Future still works at home and the best entertainment option she has is pinochle. Women playing an old Jewish men's game was Cohn's idea of radical social change.

The impossibility of disruption in both the corporate and Cohn's speculations is why they ultimately failed in the way that they did. They could not accommodate any real change in American society or values, or even a change in the nature and pace of technical advances. Utility executives did not comprehend the rising importance of environmentalism or foresee geopolitical crisis. They were shocked to find that their costs of generation could rise and doubly shocked that the kilowatt-hour price of electricity was not the only metric of importance to Americans.[10]

Then, as the 1970s rolled on, that the nation was undergoing a nearly unimaginable transition became clear. But if policymakers and the public couldn't be guided by simple demand forecasts, how were they supposed to make decisions? To displace the erroneous forecasts, energy conservation promoters needed to present an alternative way of thinking.

A group of physicists, drawn to energy problems by the oil crisis, synthesized a new way of thinking about power plants by asking a simple question: What do we need all this energy for? Their credentials turned them into incredibly effective and influential advocates for energy conservation, giving them an instrumental role in constructing—inventing—an alternative vision for the American energy system.

The most well known of these physicists is Amory Lovins, something of a child prodigy and the kind of character people like to call an *enfant terrible*. His classic *Soft Energy Paths* is probably the single most influential energy policy paper in American history not because of the details of his argument but rather because he was able to articulate a coherent point of view about what America should do at a time of tremendous turmoil.

Few people inspire more or greater divergence of opinion in the energy world than Lovins. He argued for a future based not on technological imperatives, which were really just rules of thumb aggrandized into laws, but instead on how life should be. Lovins exposed that lurking within the carefully crafted projections of 7 percent growth in electricity per year, there was a social argument, an argument about the way America should be. When he rendered the visions of the technocratic utility executives in plain English, they seemed almost ludicrous.

He ticked off what would be necessary to satisfy the sky-high projections of demand just through 1985: 900 new oil wells, 170 new coal mines, 180 new coal plants, 140 new nuclear plants, and 350 gas turbines. By the year 2000 the numbers got downright preposterous: 450 to 800 nuclear reactors, 1000 to 1,600 new coal mines, and 15 million electric automobiles. The energy industry alone would have required almost 75 percent of all the "net private domestic investment" from 1976 to 1985.[11] If we were building all that stuff, there'd be no money left over to support other businesses. When he pulled these numbers out of the infrastructure planning world and asked his fellow citizens, "Is this what you want?" many of them, including politicians, answered, "Huh. Actually, no!"

Both Lovins's supporters and his opponents have slightly exaggerated Lovins's role, perhaps. This is not to say that he hasn't been influential or is not a brilliant guy, but attacking or supporting the particularities of Lovins's style is easier than arguing against the many other physicists who became interested in energy and also started to call for more rational energy policies.

Although we don't recognize them as such, they are the unsung architects of our current system, and in 1974 most of them knew almost nothing about the energy business.

ART ROSENFELD AND THE INVASION OF THE PHYSICISTS

On a sweltering Monday in July of 1974, a remarkable month-long seminar of renegade physicists began on the finely manufactured landscape of Princeton University. Where Einstein had walked the gently curving pathways and the nation's elite learned how to pull the levers of power and money, a small group under the auspices of the American Physical Society came together to figure out how to solve the energy crisis engulfing America. Their ideas, policy prescriptions, and actions challenged the technical authority of the nation's largest energy companies and set the nation, led by California, on an energy path that would not have been possible without their contributions.

They forced redesigns of refrigerators, air conditioners, utility rate structures, assembly lines, and engineering culture. Most fundamentally, they broke the psychological link that generations had made between increased levels of fossil-fuel use and economic growth. It is almost impossible to overstate the impact that they have had in shaping the way people think about concepts like "energy efficiency" today, and yet when that summer session of the American Physical Society began, most knew little about the industrial and bureaucratic structures that they would soon be advising and critiquing.[12]

It didn't take long for the big brains gathered at Princeton to make their impact felt at the national level. Their ideas and credentials had such power that within two years, Arthur Rosenfeld of the Lawrence Berkeley National Laboratory was regularly testifying before Congress on energy-related issues. He often argued that utilities were overestimating future demand and that simple energy efficiency programs could supply Americans with all the heat, light, and power they needed—and save them money to boot.

Rosenfeld and other luminaries, like Princeton's Robert Socolow, had convened the summer session in response to the first energy shock of the postwar period, the OPEC oil embargo that led to gas rationing and a general crisis among the population. "It took us about two days to discover that the real answer was that in the U.S. energy was dirt cheap and things that are dirt cheap get treated like dirt," Rosenfeld recalled recently.[13]

Americans were shocked and dismayed, but the range of basic responses was fairly limited and straightforward: Get more oil domestically, get OPEC to call off the dogs, or simply use less oil. The OPEC-American standoff was international politics at their most intense, but it did not require the entire world to work together to solve the problem.

Most conventional energy analysts recommended simply producing more oil and building more power plants. The nation's petroleum resources, Alaska included, seemed unlikely to yield huge quantities of petroleum in perpetuity, so some suggested turning coal into liquid fuel, just as the Germans had during World War II. This thinking eventually led to the creation of the Synthetic Fuels Corporation, known as one of the bigger government boondoggles in American history.[14] But the standard analysis suggested that without more energy inputs, economic growth would be difficult to sustain. Utility executives and energy analysts argued that for twenty-five years, energy use and gross national product had walked in lockstep.

But the physicists didn't see things that way. They saw merely correlation, not causality, and they did not buy the arguments then in vogue, which called for huge numbers of new power plants or, later, establishing the Synthetic Fuels Corporation to make liquids from coal. They asked a different question, one that seems fundamental, but was largely ignored: "What is energy being used for?"[15] They went from the bottom up, not the top down.

They broke down the "energy services" humans need and want, such as space heating and cooling, light, transportation. If this is what we want, they asked, what's the minimum energy we could use to supply that? They brought back the science of thermodynamics—the interaction of heat with other forces—and tried to understand the manmade energy system like one might analyze a giant machine. Looking back to that first study, Socolow wrote in 1986,

Such an inquiry establishes thermodynamic minimum energies for the performance of tasks, against which one can compare the actual energy use. Where the discrepancies are large, one concludes, tentatively, that society has not yet been particularly clever about providing the best devices to accomplish the task, and one predicts, again tentatively, that large energy savings will become possible—with no loss of amenity—as technology evolves. Far from seeming like a guide for prediction, the 25-year constancy in the ratio of energy use to GNP struck the summer-study physicists as a testament to the sustained inattention to energy use on the part of the technological community.[16]

Rosenfeld said at that the time he began the Princeton study, "All I really knew was that the Europeans got along on about half the energy we did." He, like many particle physicists, had spent time in Europe and knew that his French and Swiss counterparts certainly were not "freezing in the dark," as the anticonservation bromide went.[17] That alone made him suspicious of the grand narratives pushed by the U.S. energy industry.

At the time, Rosenfeld would have been a quite unlikely protagonist in the American energy story. He had helped Luis Alvarez win a

Nobel Prize and aided the discovery of the empirical results that led to the formulation of quark theory. He had a comfortable and much-sought-after position at Lawrence Berkeley National Laboratory, a scientific Shangri-La, if ever there was one. He could have sat in his office perched in the Berkeley hills overlooking the San Francisco Bay, eating lunch gazing on Strawberry Canyon and carrying out the kind of particle physics to which he had devoted twenty-seven years of his adult life.[18]

But he didn't. Rosenfeld and his band of converted physicists like Lawrence Berkeley's Sam Berman and the University of Michigan's Marc Ross burst onto the national political scene with analyses that showed how to save Americans money and barrels of oil.

They were just in time.

On the day the Princeton seminar began, *Time* magazine took out an ad in the *New York Times* advertising its latest issue on the real crisis facing the country. It wasn't just the structural problems of energy, inflation, and so forth, but rather it was a problem with the nation's top tier. In blaring, block letters, the advertisement asked, "Whatever happened to leadership and where are the leaders?" Political leaders from traditional backgrounds seemed unable to provide answers to Americans' needs and pain amidst the new complexities of a globalizing world. The ad stated, "The decline of leadership is acknowledged by most Americans. Its scarcity is felt in many areas of civic responsibility; its absence is recognized in fields that formerly generated leaders."[19]

Countercultural attacks on the traditional civic, corporate, and educational institutions, widely televised, had done serious damage to the credibility of those running the country. Trust in the powers that be plummeted.[20] Just after the study in August of that summer, Richard Nixon resigned the presidency in what may have been the lowest point of twentieth-century American politics.

Who could be trusted to provide honest answers? Perhaps more importantly, what kind of methodology could be trusted? One answer to *Time*'s query—and these questions—turned out to be Art Rosenfeld.

As Rosenfeld has approached and then passed his eightieth birthday, thirty-five years into his career in energy, the current generation of environmental scholars has repeatedly feted and toasted the man whose scientific and policy breadth were only exceeded by his pragmatic will and skill in getting things done. What separates Rosenfeld from many analysts has been his willingness to get his hands dirty in both the scientific and political trenches while also maintaining a broad perspective on what does the most social good. Over the last thirty-five years, he's transformed himself from a physicist running the Bevatron particle accelerator into one of the most respected energy analysts and policymakers in the world. His accomplishments are astonishing.

He coded DOE-2, a computer program that was used to quantify the solar inputs in building design.[21] The software, which was adopted into California's building code in 1978, put solar energy on a fair analytical playing field with fossil fuel–powered HVAC systems. He created the Center for Building Science at Lawrence Berkeley, which did important research into windows that keep heat inside and compact fluorescent lightbulbs. He testified before Congress many times before entering the government in the Clinton administration as a senior adviser in the Department of Energy. In 2000 he became a California Energy Commissioner, part of the powerful board that has been one of the strongest proponents of green technology in the country.[22]

Many people have fought for what they thought was right or trained highly skilled students or conducted original research or ran institutions that changed the way the world works or served in state and national level government. But Rosenfeld did *all of that*, after working in a Nobel Prize–winning particle physics group.

Rosenfeld is gifted with the common touch. He has a way of making refrigerator efficiency and thermodynamics seem both important and simple. His innate sense of fairness also led him to seek out solutions that could benefit everyone, especially those with the least.

The self-described "epiphany" that launched Rosenfeld on this tremendous path occurred one Friday in the winter of 1973. He was working late and everyone had gone home. At the time, California burned oil for electricity generation, so he began to think about how many gallons of gasoline-equivalent his office lights used over the weekend. His calculations came out to about five gallons, half as much gas as he needed for his little Fiat. He switched off his lights and then figured that he'd turn off the lights in the other nineteen offices on his floor. Why not? He'd save all of them energy.

In the offices of his European colleagues, the procedure was simple enough: He reached in and switched off the lights. But in the offices of the Americans, there were bookcases and posters and "all sorts of damn stuff" blocking access to the light switches. His colleagues' offices became a tiny microcosm of the infrastructural problems of America. Few had ever thought about turning off the lights when they didn't need them, so the life systems they built did not allow them to have that flexibility. American society was, by design and neglect, terribly wasteful.[23] Energy usage, and even its control, had been made invisible.

Although part of the solution to that problem may be to make energy usage visible—that's certainly a major answer from the design field—Rosenfeld focused on a different strategy. If people wanted to forget all about energy, and it appeared that they did, he would simply swim upstream from where consumer purchase decisions were made and change the options they had. He would embed invisible improvements, invisible efficiency, and, in so doing, push efficiency out to every single person, not just those prone to thoughtfulness.

Perhaps the best example of Rosenfeld at work is the efficiency improvements in refrigerators. One might assume that the more expensive a refrigerator, the higher its efficiency—that higher costs meant higher performance. But when Rosenfeld and his graduate student David Goldstein actually looked at refrigerators' purchase price and efficiency, they found that they were all over the map. "[We] discovered that there was absolutely no correlation between refrigerator retail price and efficiency," Rosenfeld wrote in his autobiography.[24] That meant that they could simply outlaw the sale of the inefficient refrigerators and consumers would be "forced" to save money. "We quickly realized that if the less efficient half of the model group were deemed unfit for the market, the consumer would not perceive any change in the market range of prices or options while being forced to save an average of $350 over the 16-year service life of a refrigerator," Rosenfeld said.[25]

This would not be an operation that could be accomplished from the outside. Only the government could set across-the-board standards. As it turned out, John Holdren, then a Berkeley professor, was very familiar with Rosenfeld's work and happened to know that then-California governor Jerry Brown was going to be on campus. Holdren encouraged Rosenfeld to explain the refrigerator work to Brown over dinner at the faculty club.

At 7:45 a.m. the next morning, Brown, obviously excited, called California energy commissioner Gene Varanini and asked if Rosenfeld's numbers could possibly be right. Varanini thought they were, and within two years, new standards for refrigerators and freezers were put into effect. The California standards were eventually adopted nearly *en toto* at the federal level. Now the savings resulting from more efficient refrigerators—as compared with the 1975 models—can be calculated at $16 billion per year.[26] As Rosenfeld likes to point out, that number stacks up well with the $17 billion worth of wholesale electricity sales by all the country's nuclear power plants.

There is a graph that Rosenfeld likes to show when he gives one of his many talks. It shows electricity consumption in California and the United States since 1960. From 1960 to 1974 California follows the national trend almost exactly. But from that year until 2008, the country continues to rise while California's electricity usage stays roughly flat. In energy circles, they even call this the Rosenfeld curve or "the Rosenfeld effect."[27]

Consumers in the state of California now enjoy a situation unlike almost any other state. They pay higher than average prices for each kilowatt-hour of electricity but that energy does more useful work heating, cooling, or lighting. The net effect is that Californians actually pay smaller total energy bills than their counterparts in other states despite the higher unit costs. But that famous graph requires a bit of nuance. In his talks to technical audiences, he takes pains to note that he doesn't think California's energy policies are solely responsible for that entire discrepancy. About one-third, he says, can be attributed to the policies he helped put in place.

The famous Rosenfeld effect has received detailed scrutiny in recent years. A study by the director of Stanford's Precourt Institute for Energy Efficiency, James Sweeney, called "Deconstructing the Rosenfeld Curve" pegged the maximum contribution of California policy to 23 percent of the difference between the state and the nation.[28] The rest of the Rosenfeld effect is debatably due to what Sweeney called "structural factors" like urbanization levels, household size, commercial floor space per capita, and the inherently nicer climate in California.[29] Even so, Rosenfeld's accomplishments are still remarkable. Few people could be said to have changed so much with such little investment and so few resources.

In March 2010 dozens of scientists signed a paper proposing to name a new unit of measurement "the Rosenfeld." It would denote the energy and money saved by *not* building a five hundred–megawatt coal-fired power plant. The honorific is a symbol for how influential Rosenfeld has become and how deep his ideas have reached. One of his closest collaborators at Berkeley, John Holdren, is now President Barack Obama's science adviser. And as Amory Lovins noted in 2008, Holdren's students "have gone on to take over much of the apparatus here and abroad."

REBOUND RELATIONSHIPS AND THE GLOBE

For all his successes in the United States, Rosenfeld's methods may not be the answer for the entire globe. He has long been concerned with making things cheaper for consumers. He's fundamentally interested in the American problem of energy waste and inefficiency. Unlike younger greens, whose interest was formed during the more recent runup in oil prices and climate change concern and was tempered with blistering energy usage growth in the developing world, Rosenfeld is more interested in saving dollars than pushing any kind of low-carbon energy supply technology—whether nuclear, renewable, or otherwise.

In his autobiography he notes that the rise of climate change concern changed his priorities from "1. Money, 2. Resources, 3. Pollution to 1. Money, 2. Pollution, 3. Resources." In other words, climate change has not made Rosenfeld fundamentally rethink his position on energy efficiency: It's a good idea, and it will happen to cut carbon dioxide emissions in the United States.[30]

It shouldn't take anything away from Rosenfeld's legacy as the "godfather of efficiency,"[31] but there are persistent and pesky questions about whether or not efficiency will actually reduce energy consumption and thus carbon dioxide emissions in the long run. A gaggle of analysts have questioned whether a "rebound" effect eats up most or all of the energy initially saved.

In the simplest case, there can be a direct rebound effect: If your car gets thirty miles to the gallon rather than fifteen, you might drive more.

The direct impact of such changes is the easiest to measure and proba- bly to manage, too. An intensive 2009 review of the empirical evidence found that this direct rebound effect occurs for probably less than 30 percent for household services in developed countries.[32] From this number, Rosenfeld's main criterion seems to be satisfied. Consumers will obviously save money and get more energy services. Yet even this moderate rebound effect is not fully factored into the calculations by esteemed bodies like the International Energy Agency when they con- sider efficiency as an emissions reduction strategy.

But there is a deeper and more global problem with energy effi- ciency as a carbon reduction strategy. If efficiency measures in the de- veloped countries (perhaps accompanied by a carbon tax) reduce fossil-fuel usage in those countries, economists would say that the price of those fuels could drop. Petroleum, natural gas, and coal all trade at pretty close to the same price globally, so some demand destruction in the United States may make using fossil fuels more attractive to a grow- ing country like Indonesia or China. Some critics have suggested, fol- lowing the early thought of the nineteenth-century British economist William Jevons, that by lowering prices, efficiency may well cause in- creased, not decreased, consumption. Economists call this backfire.[33]

The most obvious and suggestive evidence for the Jevons Paradox is that although the amount of energy per dollar of GDP keeps dropping, the absolute amount of consumption keeps rising. This is true the world over—and it's not simply a function of greater numbers of hu- mans. Per capita energy consumption, the vast majority of it derived from burning carbon-based fuels, continues to grow. One study looked at how this happens in California. They found that American home sizes just kept growing and growing. The efficiency gains of better HVAC systems have been completely overwhelmed by the growing vol- ume of air that has to be heated and cooled. In 1950 the average person lived in 286 square feet of space. That's a little less than a couple of parking spaces. In 2000, however, the average floor area per capita had risen to 847 square feet. That's only a little less than the total square footage of an average 1950 home.[34]

As such, the role of energy efficiency has become linked with some of the most contentious debates in energy economics. Whether energy or energy services drive economic growth or whether economic growth causes greater energy consumption still remains unclear. The nature of the link between energy consumption and economic growth remains disputed, along with just about everything else that could provide clarity on how effective energy efficiency may be in reducing general energy consumption and, specifically, carbon emissions.[35]

The most sophisticated analyses are now trying to determine which energy efficiency improvements are most likely to have large rebound effects. Steve Sorrell of the Sussex Energy Group has split up the big efficiency targets into two large groups. On the one hand, efficiency gains in developing countries, with general purpose technologies like internal combustion engines, and in energy-intensive sectors appear likely to be susceptible to major rebound effects. On the other hand, efficiency improvements in developed countries, specifically in households, may lead to actual energy consumption drops.[36]

Perhaps the best framing for energy efficiency's role in the greentech future comes from Dan Adler, president of the California Clean Energy Fund. "Efficiency is how we're going to pay for all of this other stuff," he reflected.[37] Any energy supply—some form of renewable or nuclear—that is brought onto the U.S. grid so that the country can shut off its coal plants will probably cost a little bit more than coal. Consequently, following the California model of using invisible efficiency improvements to offset the price increases of new energy types could be a political necessity.

All of these studies necessarily rely on data from the past, when energy supplies were not constrained. It remains to be seen how the peak of global oil production will play out, but Taoyuan Wei of Oslo's Center for International Climate and Environmental Research pointed out in a 2009 paper that nearly all the rebound studies pretend that there is an infinite amount of energy supply.[38] That is to say, the new studies on the rebound effect focus just on the demand side of the global equation. In a kind of sad engineering poetry, the new research almost perfectly inverts the supply-obsession error that Rosenfeld's generation helped to correct.

This means that if the security of the energy supply enters a period of remarkable deterioration, past rebound effects may not be a good guide to future rebound effects. Although it may have made sense at some point in the United States to continue moving farther and farther outside urban cores and using increased fuel efficiency savings to drive more miles, those days may be coming to an end. Even coal reserves, which had been thought likely to last for well over the next one hundred years, are getting a fresh look and revised downward by some analysts.[39] And efficiency savings that may have been converted into greater consumption in the past may not have the same effect in a future of rising energy prices.

Conversely, if new carbon-based energy supplies are easily tapped and the world does not put an across-the-board limit on their extraction, then it seems likely that no amount of low-carbon energy or efficiency, no matter how cheap it is, will keep that carbon in the ground. If the global North switches to renewables or nuclear en masse, doing so could very well drop the price of fossil fuels, thus leading to an increase in global consumption by poor countries. This may be a global social good: Vaclav Smil has calculated that real and quantifiable good stuff happens as a country reaches closer to an energy supply of about 110 gigajoules per person.[40] Billions of people are nowhere close, while Americans and Europeans far exceed the threshold. Purely from a climate perspective, however, it doesn't matter who burns the fuel.

Without global coordination and a historically unprecedented deployment of low-carbon power sources, the least savory effects of deranging the atmosphere will be felt. The only options we will be left with are the direct removal of carbon dioxide from the air and its subsequent burial in the ground as advocated by Columbia University's Klaus Lackner or massive geoengineering experiments that will have unknown and unknowable consequences. Given the ineffectiveness of current world deliberators about a carbon-capping treaty, it's not hard to see why some scientists are quietly but firmly advocating that geoengineering must be understood so that we know whether or not we have a backstop.[41]

If that does happen, there will be a curious resonance between the world imagined by Victor Cohn and the one that humans end up living in: The globe will become one huge managed climate bubble, "a terrarium" in the words of journalist Eli Kintisch. Cohn imagined,

Man in 1999 made weather on a vast scale, or at least changed it. TV scanned distant clouds, and electronic brains analyzed signals from chains of robot observers all over the world. A group of unmanned satellites or space stations, circling the earth at a height of 500 miles, observed the atmosphere over the entire globe, and relayed what their TV cameras saw to scientists down on earth. A federal agency, established in 1971, then made United States control decisions. In mountains, weather makers built deep snow packs, and a 1999 report said: "We are forming new glaciers now, and may expect more summer runoff than ever."[42]

But what for Cohn was a hopeful future is for us a worst-case scenario.

chapter 18

Transcendentalism

I N THE SPRING OF 1975 Jim DeKorne welcomed the photographer Jon Naar onto his family's one-acre homestead in El Rito, New Mexico. They toured the grounds, passing the "survival greenhouse" and its aquaculture pond, the house heated largely by the sun, and into a small outbuilding that housed the batteries for a Jacob Wind Turbine of a design more than fifty years old.[1]

The building was small and built entirely from wood. Skinny logs formed the outside of the building, simple boards the inside. A half-sized door allowed entry into the control room. On a shelf, a bank of large "Heavy Duty" batteries ran up to the wind machine's electricity generator. The only adornment was a *New Yorker* cartoon tacked on the wall.

DeKorne wore a long, untrimmed beard that called to mind both Fidel Castro and Allen Ginsberg. A herringbone cabbie hat was pulled low on his forehead, nearly to his thick-rimmed glasses. The former schoolteacher and English master's degree holder looked uneasily into the camera with his finger pointing at his energy storage system.[2] The wind-powered batteries provided electricity for a pump that circulated water from a flat-plate solar collector on the roof to the fish pond. The aquaculture tank worked like a "heat battery" for storing solar energy.[3]

By that time DeKorne was well known to the small group of solar enthusiasts who had sprung up around the country. He had published a series of articles in the back-to-the-land magazine *Mother Earth News*, describing his various efforts to build a totally self-sufficient existence out on the New Mexico plateau. He had created the Walden Foundation to publish a book, *The Survival Greenhouse*, describing an underground system for hydroponic agriculture.[4] He and his wife, Elizabeth, also decided to create an "eco-system" that was "capable of supplementing the diet of a small family" while staying "consistent with ecological reality—only natural, non-polluting systems could be considered." Their secondary considerations were "maximum yields, minimum waste, and reasonable ease of operation and maintenance."[5]

The raw material was a one-acre plot of land they purchased in 1970. Over the next five years, it became a roughly self-sufficient homestead. They had the greenhouse and a sauna, a root cellar, and a pigpen. The family's big idea was to sink the greenhouse four feet below the ground, allowing them to use the soil as an insulator. This "grow hole" was part of a looping ecosystem composed of earthworms, plants, and rabbits. DeKorne had stocked a solar-heated fish tank with bluegill sunfish from the closest pond and caught them "with barbless hooks," but the fish required more food than they provided, so they eventually stopped farming them.[6]

The DeKornes nearly dropped off all the nation's energy grids. Solar energy of one type or another provided most of their food, heat, light, and power. Although they found their lives fulfilling, or so they told outsiders, they lived a spartan existence with few of the things that most Americans spent their time acquiring. What were they after? DeKorne indicated that by changing his energy usage and lifestyle, he would be making change in the world. Essentially, in a political climate characterized by a feeling of helplessness, the back-to-the-landers chose to drop out. "If you want to change the world, change your own life," he told Naar.[7]

DeKorne wanted to live with the rhythms and energies of nature—and without the strictures of civilization. The quest, though it had social implications, was deeply personal. DeKorne had turned his back on the consumer culture to live "a Thoreau-like life with a wife and two kids." Quite intentionally, he and his cohort recalled the transcendentalism of the 1840s.

PUTTING THE AXE TO THE ROOT OF SOCIETY'S EVIL

The 1840s in America were a time of tremendous technological change. In 1830 there were twenty-three miles of railroads. Twenty years later, there were nine thousand. The first telegraph message was sent in 1844, and suddenly information could travel faster than any animal or machine. At the same time, more and more white men could vote as suffrage began its long, slow spread. The banking system grew larger and more powerful. By 1860 there were 1,500 banks, each issuing its own notes and sending money flying around the country.[8]

This was the world of Henry David Thoreau and his short stint at Walden Pond living on Ralph Waldo Emerson's property in his homemade cabin. The American continent was in the process of becoming networked: Information, transportation, citizens, and goods were beginning to flow easily from one place to another. More people and things covered more distance than ever before, and the difficulty of getting almost anything from one point to another was decreasing.

As early as 1828 James Kirke Paulding could write a guide to New York for tourists called *The New Mirror for Travelers and Guide to the Springs*. "All ages and sexes are to be found on the wing, in perpetual motion from place to place. Little babies are seen crying their way in steam boats, whose cabins are like so many nurseries," Paulding wrote.[9] Thoreau might have recoiled from this world, but many found it more exciting than the world into which they had been born. Conversely, the dislocations of the time were also deeply disturbing. Dozens of groups of thinkers and activists sprang into the American public sphere. "In the 1840s there were dozens of reforms competing for the honor of naming the root of society's evil and putting the axe to it—abolitionism, Grahamism, phrenology, prison and asylum reform, temperance, pacifism, compulsory education, women's rights, land reform, workingmen's associations, and so on," wrote historian Taylor Stoehr.[10]

The Transcendentalist response to all the changes for good and ill in American society was, largely, to focus on the self to the exclusion of societal concerns. In particular, DeKorne's hero, Thoreau, was prone to drop out. Even the deepest and most generous examination of the Transcendalists as a social movement excluded Thoreau because he was "interested in social reform only by way of personal example."[11]

Whereas industrialists and utopians like John Etzler imagined that machinery could transform the world, Thoreau believed that social change could be accomplished only through internal changes. In reviewing Etzler's green-tech utopian manifesto, *The Paradise Within the*

Reach of All Men, Without Labour, By Powers of Nature and Machinery, Thoreau took issue with Etzler's vision of a world in which everything could be done by the "motion of the hand at some crank." Thoreau wrote, "But there is a certain divine energy in every man, but sparingly employed as yet, which may be called the crank within,—the crank after all,—the prime mover in all machinery,—quite indispensable to all work. Would that we might get our hands on its handle!"[12]

As we saw in the introduction to this book, for Thoreau the "crank within" was all that mattered. Whereas men had always had a hard time reaching that spiritual sense, the new machines, monetary systems, and general complexity of industrial society had estranged humans from their purer natures. "The laboring man has not leisure for a true integrity day by day; he cannot afford to sustain the manliest relations to men; his labor would be depreciated in the market," Thoreau argued in *Walden.* "He has no time to be anything but a machine."[13]

Thoreau's answer to the "fool's life" of Concord—the "seeming fate, commonly called necessity"—was to repair to a solitary hut by Walden Pond. Leaving other humans behind would allow him to focus on himself and develop his consciousness by communing with what was natural. "I go and come with a strange liberty in Nature, a part of herself," he wrote. To him, she became "sweet and beneficent society" as "every little pine needle expanded and swelled with sympathy." Nature's rhythms filled his life, and if it rains and keeps Thoreau in his house, it's no bother because "being good for the grass, it would be good for me." Thus, the year's temperature swings and weather events condition him to know, morally, what was natural and right.

This attitude of natural revelation is what still draws us to Thoreau. It seems he can access a reservoir of natural righteousness and pipe it directly to us. It is no accident that DeKorne's publishing ventures went out under the imprimatur of his own devising: Walden.

"THE SUN RENEWS US, IN AN ALMOST RELIGIOUS WAY"

The photographer Naar included DeKorne's home in his book, one that followed a collaboration with Norman Mailer on graffiti culture in New York and Norma Skurka, a *New York Times* writer. The book was named *Design for a Limited Planet,* and it explored the range of hippies and architects who were abandoning their city lives and heading out to places like northern New Mexico to build a new type of home that used the sun instead of fossil fuels for space and water heating. Bouncing around the countryside, Naar "captured a key moment in the history of the solar architecture in the United States, namely the few years during which it was no longer the scientific curiosity it had been prior to 1973, and not yet the somewhat larger phenomenon it would become after 1977 when Carter's incentive policies were implemented."[14]

The people who were drawn to solar energy then believed in the kind of self-sufficiency and value for nature that Thoreau espoused. But their unique contribution to American thought was the idea that changing the way they used energy was changing them *as people*. Just as Thoreau became "a part of herself," many of the solar advocates held the rather remarkable belief that changing the space heating arrangements used to regulate the temperature of their dwellings was bringing them something like spiritual enlightenment.

In *Design for a Limited Planet*, homeowner after homeowner spoke to the spiritual changes they underwent while living in a home that was connected with the earth's systems. One said that making his home less dependent on a utility company "is primal and gives you a feeling of control over your destiny."[15] Another, Paul Davis, said his house gave him "a sense of being more in touch with nature, not cut off from it. A house like this becomes an extension of your body." Davis didn't even like modern control mechanisms, instead controlling his ventilation system with "a pull chain strung with colored beads" because that felt "a lot nicer than throwing a switch or fiddling with a dial."[16] A Taos resident living in a home patterned on Michael Reynolds's famed early "earthships" and built partially out of beer cans called his "the happiest house we have ever lived in."[17]

Perhaps Junius Eddy, whose Rhode Island home was designed by the ubiquitous solar architect Travis Price, summed up the zeitgeist into which these solar pioneers were plugged:

It is not just the financial savings. We grow more in awe of the tenuous hold our lives have on this small planet, more convinced that the sun renews us, in an almost religious way. It has made us profoundly grateful that the sun is up there, the center of our universe, warming us up and keeping us alive. That atavistic sense of the elements that early man knew and felt has become part of our lives.[18]

Thus, there was something more fundamentally human about living in a solar house than in a regular house. "Living in a solar house is a whole new awareness, another dimension," said Karen Terry, another New Mexico homeowner. Terry's home cost one-third more than she'd planned and the temperature inside the house swung 15 to 20 degrees during the day, but she said, "It's the morality of the design of the house that's so important."[19]

Of course, when living out in the middle of nowhere, agitating for the kinds of massive social change that had marked the '50s and '60s is difficult. But like Thoreau, that the solar home owners could lead by example was certainly possible. Recent scholars have noted that there was a well-established path that could be trodden between various solar residences in the Southwest.[20] DeKorne's homestead received "hundreds of visitors each year." So many people dropped by that the DeKornes had to request that would-be gawkers request an appointment in writing in advance.[21]

There was popular appeal to the natural ideal expressed by the *Design for a Limited Planet* homesteaders; after all, the book sold more than 100,000 copies.[22] In a country filled with new, all-electric suburbs, the attitudes of the people in the book were delightfully countercultural without being too militant. Homeowners like Eddy and DeKorne were not trying to change anyone else's lives but instead choosing their own. In that, the book's subjects were very much with the time: The 1970s saw a tremendous rise in lifestyle activism. The popular writers of the day, like Thoreau and Amos Alcott before them, provided the philosophical backing for the idea that *doing something with your life* was more important than traditional organizing or political protest.

Charles Reich, author of *The Greening of America*, wrote that a lifestyle choice *was* the most effective form of activism because the rising generation, who he described as having "Consciousness III," were looking to make deeper change than that attained by traditional politics. Reich wrote,

But if we think of all that is now challenged—the nature of education, the very validity of institutionalism and the legal system, the nature and purposes of work, the course of man's dealing with the environment, the relationship of self to technology and society—we can see that the present transformation goes beyond anything in modern history. Beside it, a mere revolution, such as the French or the Russian seems inconsequential—a shift in the base of power.[23]

The mechanism for the change was to be "choosing a new life-style." Young dropouts out on the roads and in the solar hot tubs of the country were educating themselves, not just messing around. They were being transformed spiritually to prepare them to lead the new society, Reich argued. "A fundamental object of this education we have described is transcendence, or personal liberation. It is liberation that is both personal and communal, as escape from the limits fixed by custom and society, in pursuit of something better and higher," Reich continued. "It is epitomized in the concept of 'choosing a life.'"[24]

Reich, who preached a revolution of the young, might as well have quoted *Walden*. The just-about-thirty Thoreau of the Walden years thought that "the mass of men lead lives of quiet desperation." People might have thought they were choosing their "means of life," but really they were simply bound by their elders' prejudices. "What old people say you cannot do you try and find that you can. . . . Practically, the old have no very important advice to give to the young, their own experience has been so partial," Thoreau observed. "Here is life, an experiment to a great extent untried by me."[25]

But for all of Thoreau's conviction and his willingness to choose a new lifestyle, at least temporarily, his writing did little to slow the technological deepening of American society. Americans sought an inner moral strength that might change the world around them, but whatever happened internally, the world kept on keeping on. Everything he opposed—mechanical progress, the conception of nature as a storehouse for man, and so forth—became only more entrenched in society. "Emerson and Thoreau had no discernible effect on the dominant capitalist culture of their time," political scientist Martin Schiff noted in his

1973 essay, "Neo-Transcendentalism in the New Left Counter-Culture."[26] Perhaps Thoreau's very inability to reshape industrial society is what has kept him relevant for so long. The issues he identified in our relationships with nonhuman environments remain as troubled and complex as ever.

The solar transcendentalists thought that finding prefigurings of their ideas in a famously American group of thinkers gave them a better intellectual heritage, but Schiff argued this very resemblance to the Transcendentalists, far from rooting them in the mainstream of American culture, probably doomed them to failure. "This relationship between New Left counter-culture, credited by its supporters with messianic characteristics for national salvation, and a nineteenth-century utopianism that made no significant impact on the unfolding of history raises many questions about the credibility and future of the counter-culture," he argued.[27]

Perhaps comparing eras in this way is a flimsy enterprise, but the years following World War II and the 1840s stand out as times marked by rapid, radical change technologically and socially. There was explosive growth of industry, societal upheaval, and the extension of political rights to far broader classes of people.

Between 1950 and 1975 energy usage in America doubled. Oil poured into the system as Americans drove more and larger cars farther. Plastics replaced natural fabrics. Petroleum-based chemical use grew in farming; nitrogen fertilizer and pesticide applications per acre grew nearly tenfold. Meanwhile the Vietnam War and its chemical defoliants and napalm provided a dark shadow for the shiny new wonder products.[28]

The agricultural, infrastructural, and cultural worlds into which baby boomers were born had been transformed before they would hit thirty years old. They knew a big part of those changes were due to the changes wrought by increased energy usage. Every kind of commercial network was growing larger while social networks seemed to be growing thinner. Books like Rachel Carson's *Silent Spring* and TV shows like *Our Vanishing Wilderness* had shown the overwhelming impact humans were having on each other and life more broadly.[29] Human beings built rockets to carry themselves to the moon, yet, as the Gil Scott Herron poem pointed out, "I can't pay no doctor bill / but Whitey's on the moon." Technology was big and seemingly doing as much harm as good.

Something fundamental was flipping in American culture. The modernist dream of technology making the world a better place was falling apart. Technology seemed to be impeding social progress, not aiding it. The future no longer seemed to promise something better than the past.

As in the 1840s, all the mechanization and energy use and corporatism had some people feeling like American society had gone off the rails. There was just something fundamentally wrong with living the way Americans did. The very energy surpluses, the very economy that their parents and grandparents had built seemed like a perversion of what being a human was supposed to be. Political ecologists like Barry Commoner rode this wave of sentiment to popular fame. Environmental problems, like the smoke that choked nineteenth-century cities, were simply canaries in the societal coal mine. The energy crisis was, to Commoner, "a symptom of a deep and dangerous fault in our economic system."[30]

To a well-read intellectual like DeKorne, the parallels between the 1840s and the tumultuous times he grew up in were obvious. He wrote in a 2009 mini-memoir that

> the parallels between the themes of the 1960s and the transcendentalist era of the 1840s became ever more striking—it was history repeating itself on another octave. A kind of escape was what the transcendentalists were selling over a hundred years before, only they called it 'transcendence,' and transcendence had great appeal in the 1960s, even more than it did in the days of Emerson and Thoreau. Apparently, the transcendental ideal is one of those relief valves that blows off every so often when the pressure becomes unbearable in our American boiler.[31]

The solar transcendentalists drew from a wide range of sources for their philosophies. There were the deep American antitechnological roots of Thoreau and there were the popular philosophers of the day like Reich. Anarchist writer Murray Bookchin's human-scale technology ideas stewed with economist E. F. Schumacher's Small-Is-Beautiful mantra. The apocalyptomania of Paul Ehrlich's *The Population Bomb* cross-pollinated with the home-spun, quirky optimism of Stewart Brand's *The Whole Earth Catalog*. Farrington Daniels's buttoned down solar research mingled with Buckminster Fuller's all-night bull sessions about energy transformations. In magazines like *Rain*, Amory Lovins's nuanced and technical "soft energy path" arguments shared space with primers on how to build composting toilets, grow organic kale in the city, and mend your blue jeans.

Over it all, the imaginary contrails of ballistic missiles flying between Moscow and Washington made the ideological differences of the world's superpowers seem secondary to their union in adhering to the industrial system with its dependence on science, machinery, and high technology. Heading back to the land, singly or in communes, provided a way out of the Communist-Capitalist divide, allowing for a rejection of government and authority generally, regardless of their position on the control of the means of production. To hell with the both of you, back-to-landers said.

"Communism is essentially 'people's capitalism'—and, as an economic system, it is not inherently more 'ecological' or less damaging to natural systems than capitalism is," DeKorne wrote in *The Survival Greenhouse*. "Neither system is adequate to cope with the real problems we face."[32]

The political system seemed hopeless. The world was on the verge of destruction and the governments that should have kept it alive were asleep at the switch. In the face of all that, perhaps concentrating on developing one's consciousness while vainly hoping that the world would see that a more ecological way of life was possible or preferable made sense.

For a world of local environmentalism, individual action made sense because it's comparatively easy. Serious change can be accomplished without the difficulties of large-scale organizing or global politics. The story of Bolinas, California, is a classic example of a successful small-scale effort to make environmental change. A group of dedicated, ecology-minded residents banded together to fight the introduction of a sewer system. It might seem an odd target, but led by *Whole Earth Catalog* contributor Peter Warshall, they realized that the

Small Nebraska towns were blanketed in water-pumping windmills around the turn of the twentieth century.

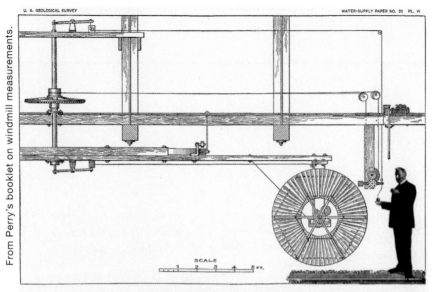

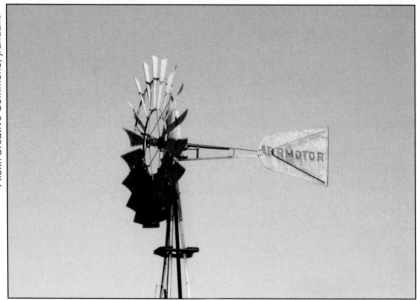

TOP: Thomas Perry's windmill testing apparatus. BOTTOM: Many Aermotor windmills, patterned on Thomas Perry's original design, remain standing today.

TOP: We can see a Halladay standard windmill in this photo. It was the first self-regulating "American-style" windmill. BOTTOM: A homemade windmill still standing in Nebraska. Thousands like it were constructed in the late nineteenth century.

The Duffy Wave Motor was one of many that never succeeded.

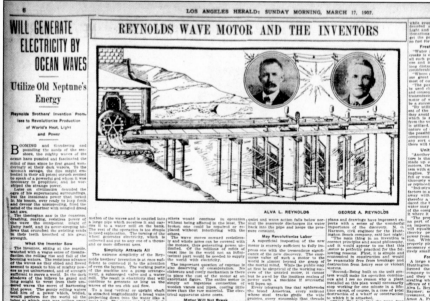

The Reynolds Wave Motor was regularly hyped by the *Los Angeles Herald*, whose editor happened to be a shareholder.

🜨 Solves Problem of Harnessing Ocean Waves 🜨

THE problem of extracting heat, light and power from the ocean's immeasurable energy has been solved by Alva L. Reynolds of Los Angeles and Huntington Beach. "We have made a capital demonstration at our plant at Huntington Beach," said Mr. Reynolds. "We are making electricity every day, and hundreds of persons are calling at our pier to see the motors work, and they go away convinced that this age-long problem has been solved. Our two motors develop more power than we anticipated. Within a short time more units will be added, and then we shall be ready to put our electric energy on the market."

Asked to give in detail an explanation of how his motor was operated, Mr. Reynolds said: "I will give you a brief review of the wave motors that have been attempted and then show how we have overcome the obstacles which have baffled other inventors.

"The history of wave motor inventions shows about 175 different schemes for utilizing wave power, each representing as many different notions, not scientific knowledge, about wave action. These inventions may be classified as follows: First, those attempting to use the vertical action of the waves, and strange as it may seem in the light of all the scientific information at hand, this class includes 95 per cent of all patents issued from the patent office; second, the pendulum idea, and, third, the vane principle.

"All the inventions of the first class employed a float or barge or a combination of these, by means of which it has been attempted to transmit, through an intricate mechanism of gears, wheels, ratchets, clutches and air pumps the rising and falling movement of the waves. Had the vertical action of the waves been the only factor to be reckoned with these would-be inventors would long ago have solved the wave power problem, for there is great power in that action. But the horizontal component of wave power has been and is a destructive element to every resisting surface, and

no float or barge of sufficient area to be of any practical utility in transmitting wave motion can ever be securely anchored to resist the horizontal component of wave power. As proof of this statement is the fact that every float or wave motor involving the float principle installed in the ocean has been destroyed in a short time.

"The second class of wave motors includes the pendulum and pendulum float, the latter being a modification of the float principle. Only six or seven of these inventions have been presented to the public, and in every instance where a test was made in the ocean the device was wrecked, because of its inability to resist the attack of the waves. Another objection to, and thus far an insurmountable difficulty in the utility of, the pendulum principle is the enormous cost of wharf construction made necessary by the very principle involved. Whether or not mechanical difficulties can be overcome and whether or not a pendulum wave motor can be installed which will be able to resist the destructive power of the waves remains to be demonstrated. All scientific data at hand points to the probable failure of all efforts in this direction so far as pendulums are concerned.

"The third class of wave motors, of which only one has been tested in the ocean and successfully withstood the severe storms of winter, is constructed on the vane principle and utilizes only the horizontal action of the waves and offers practically no resistance to such action.

"Mr. Reynolds has profited by the many failures of other inventors by studying the science of wave action and applying its known principles to the solution of the hitherto difficult wave power problem. That they have solved this problem seems apparent from the results of the recent demonstration at Huntington Beach, where two of these units were installed and withstood for several weeks the severest storm known on this coast for many years."

In a letter to Mr. Reynolds Capt.

WAVE MOTOR WHARF AT HUNTINGTON BEACH FROM SHORE SIDE

Amos A. Fries, United States engineer in charge of the great work at San Pedro harbor, wrote:

"By utilizing the horizontal component of the wave instead of the vertical, you avoid the use of any sort of float, which is always liable to destruction from the very horizontal component that you utilize. On the other hand, the very nature of your motor almost precludes the possibility of injury by the vertical component of the waves. In fact, where any sort of wharf or pier will stand there will your motor stand.

"Another excellent and most desirable feature is that every power plant will be made up of a number of your simple motors, thereby making it a unit system, with all the advantages which that implies. Thus, one or more units may fail or wear out and the plant run

continuously while they are being repaired or replaced.

"Indestructibility is one of the chief factors in any wave motor, and by doing away with the float and substituting therefore a vane you have reduced the breakage to a minimum. The water may chase the vane to all points of the compass, but it can hardly catch it where it can be broken.

"Your proposed storage system is an important part of the plant, as it will equalize the wave energy and the eby produce a steady output of power from the waves' intermittent action. With a large number of units properly placed a reservoir may not be necessary, or, at most, only a comparatively small one.

"Taken altogether, your plans appear to me to be practicable, and your motor ought to be a success, though only a trial will demonstrate what it is capable of doing."

INTERIOR OF FIRST POWER HOUSE ON WAVE MOTOR WHARF
HUNTINGTON BEACH

PICTURE SHOWS TWO PUMPS AT TOP OF WAVE MOTORS, THE PIPING AND PRESSURE TANK. ALVA REYNOLDS, THE INVENTOR

Like several other wave motors, the Reynolds wave machine made it all the way to prototype construction before vanishing from California history.

TOP: One of the last extant examples of the Day and Night Solar Hot Water Heater, which was installed at Scotty's Castle in the Mojave Desert. BOTTOM: A generator from the Folsom Dam powerhouse, part of the project that began the electrification of California.

The Lady Hunter Oil Well, an example of an early well flowing uncontrolled.

TOP LEFT: The Smith-Putnam wind machine, the very first megawatt-scale turbine, constructed before World War II. TOP RIGHT: A photograph from the construction of the Smith-Putnam, showing what appears to be a man holding a huge wrench. BOTTOM: Another photograph from the construction of the Smith-Putnam, showing the construction of the rotor.

The cover of the Stewart Brand–led magazine, the *Last Whole Earth Catalog*.

LEFT: An old-style windmill used to pump water near a New Mexico eco-settlement in the early 1970s. RIGHT: A handmade windmill at an eco-settlement in the early 1970s.

TOP: A solar home demonstration project funded by the National Science Foundation. Note the collectors on the roofs. BOTTOM: Physicist Art Rosenfeld became an influential leader in rational thinking about energy in the 1970s.

TOP: An early solar panel hooked up to run a television by NASA scientists. BOTTOM: Steve Baer, the pioneering founder of the solar company Zomeworks.

TOP: The Solar Energy Research Institute's entry in a local parade. It reflects unhappiness with prospective Reagan-era cuts to the outfit's budget. BOTTOM: Aden and Marjorie Meinel's prototype parabolic solar collector. That's their son Steve.

TOP: These are the first two large-scale solar plants ever built, Solar Electric Generating Stations I and II. BOTTOM: A SEGS parabolic mirror up close.

TOP: This is the site on which the massive Ivanpah solar plant will be built.
BOTTOM: Eventually the Ivanpah solar plant will look like this rendering.

From the air, this rendering shows how much of the desert valley the plant will occupy.

TOP: eSolar's first test plant in Lancaster, California. BOTTOM: The eSolar plant employs the company's distinctive small, flat mirrors.

waste infrastructure was a key pinchpoint that could stop future development in the area. "If you put in a large-volume wastewater pipe, that allows development. If you constrict the size of the pie or the extent of the infrastructure octopus, you can't have development except for septic tanks," Warshall recalled.[33] The Bolinas gang was chronicled in the film *The Town That Fought to Save Itself.*[34] To this day, the city remains a beautiful place.

But what if there is a global problem that requires nearly all the world's citizens to work together? What happens if we need to cut global carbon emissions by something like 80 percent? What happens if, to preserve the biodiversity of the world, we need to beat existing technologies in countries without democratically elected governments? Even with the same ecologically sane frame of reference, the most impactful actions of the 1970s and this decade may be quite different.

One MIT study calculated that no American, no matter how abstemious their lifestyle, could reduce their carbon footprint below 8.5 tons or their total energy usage below 130 gigajoules.[35] Even if you beg for vegetarian scraps from town to town, societal things like roads, police service, fire departments, libraries, the courts, and the military are all carried out on your behalf. And those things require energy. To reduce the carbon footprint of the country—which is more than twice the global average—society as a whole has to change.

For that to happen, the presentation of a real alternative is necessary. In truth, the solar transcendentalists were too caught up with the Bomb-induced apocalyptophilia of the time to present such a unified vision of a different society. Instead of organizing politically or socially, they were learning the "survival" skills that they almost seemed to darkly hope they would need in the postcivilization future. DeKorne held that

It is not, however, within the scope of this book to explore the almost insurmountable problems to be solved before our society could be expected to function within an ecologically sane frame of reference. Indeed, it is the author's opinion that the last meaningful chance for a societal change in this direction was passed sometime just before World War II. At any rate, survival is now an individual responsibility, and it is to individual solutions that this book addresses itself.[36]

However, most solar transcendentalists drifted away from their intense stances on energy. Some returned to "straight society" whereas others looked for transcendence without the sun. Jim DeKorne moved on to "life's greatest challenge: the soul's Gnostic commitment to the Great Work of transformation—the impossibly perilous journey through the infinite maze of hyperspace." He wrote a book on psychedelic shamanism and moved to Hawaii.[37] By 1979 one contemporary historian observed that "Numerous 1960s activists had moved from protest politics to self-awareness." Though some remained active fighting nuclear power, increasingly "looking inward seemed more fulfilling than changing the world."[38] Societal change by individual example had given way to a broader navel-gazing that left energy out of the picture. By the time Bill Clinton was elected president, it seemed all that remained of the solar transcendentalist movement was the popular memory of lukewarm solar hot water–heated showers and a few odd pieces of architecture.

Their legacy, however, is greater than it might appear.

chapter 19

Tools

VICE-GRIP PLIERS MAY SEEM a lowly object for veneration, but for the emerging group of appropriate technologists, they were material poetry. Stewart Brand's description of them in *The Next Whole Earth Catalog* could be read with line breaks as a paean to timelessness:

> A locking plier wrench
> which grips and holds
> with the size of the grip
> adjustable.
> Hold things hard without muscle,
> hold things together,
> hold things with just the right pressure.
> Many's the busted faucet you see
> with a vise-grip biting it
> as temporary handle.[1]

The vice-grip plier was a tool that let a person grab hold of the world the way it should be held. They could be adjusted to hold the world together, but without the sort of overbearing associated with modern technology. They were perfect for fixing the plumbing of the world. There was no improving on them. They were just right.

Steve Baer, a frequent *Whole Earth* contributor, loved them because they were objects that were not dependent on the infrastructure of the twentieth century. They might have been a fairly recent invention, but they *could* have been invented long ago. They could circumvent the regrettably immoral system of industrial production. "How exciting it is for someone today to come upon designs which answer needs that have existed for a long time, but which could have been explained to and implemented by someone back in 1850!—something someone could have produced in an existing shop then!" Baer wrote.[2]

His interest is not atavistic. He was not interested in using the stuff of olden times but rather creating stuff that solved problems humans have always and will always have. Explicitly contrasting them with computers, Baer wrote, "Another recent tool—the vise grip—could have been made and appreciated 100 years ago after a few moments' demonstration and explanation. How deep into history could you carry an idea and have those who saw it say, 'Of course, why didn't I think of that?'"[3]

For Baer and many appropriate technologists, tools were the permanent technologies. They were the perfect expression of a solution to a material problem—and they were measured by the length of their use, not by their newness. In answer to the question historian Carroll Pursell raised about what, exactly, appropriate technology was appropriate for or to, the answer was: all humans, regardless of their position on the gradient of time. The attitude goes right to the heart of the difference between "tool-freaks" and technologists. Tool makers were after timeless things; technologists were after new things made for the perpetual right-now.

Discussion of tools and "access to tools," the goal of the *Whole Earth Catalog*, dominated thinking about appropriate technology. The new field was a unique synthesis of "unbridled technological optimism" with the desire to create systems that would not enthusiastically destroy ecological systems. Even technology-phobic communards "enthusiastically embraced" tools. "What was so appealing about *Whole Earth* for intentional community builders was the blending of the primitive and technological, the time-tested traditions alongside the new wave," wrote historian Andrew Kirk, author of *Counterculture Green*.[4]

James T. Baldwin, *Whole Earth* editor who was one of the most important practitioners of appropriate technology, explicitly argued against any sort of rush to return to old practices, lamenting that so many people began to turn away from new stuff and ideas. In the introduction to the book he coedited with Stewart Brand, *Soft-tech*, Baldwin wrote,

Technological excess begat antitechnological excess. . . . There are people who champion a return to the ways of our forefathers, but this has not turned out to be much of an answer. Our grandparents were in many ways worse than ourselves. They saw forests as endless and topsoil beyond measure. . . . There are lessons in the past, but we shouldn't and can't go back.[5]

What direction should American society take, then, if not backward to our forefathers nor forward into the techno-future? Baldwin and Brand offered their own take on appropriate technology, hoping "alive, resilient, adaptive, maybe even lovable"[6] tools would beat out the old, ossifying power plants. The goal was to make technology more natural, but others had their own formulations. E. F. Schumacher, for example, who was incredibly influential, railed against "ever bigger machines, entailing ever bigger concentrations of economic power and exerting ever greater violence against the environment." Instead, he argued, "wisdom demands a new orientation of science and technology towards the organic, the gentle, the non-violent, the elegant and beautiful."[7]

Schumacher added another requirement to this new appropriate technology: "suitability for small-scale application." Drawing on the ideas of political scientist Leopold Kohr, he wrote that "small-scale operations, no matter how numerous, are always less likely to be harmful to the natural environment than large-scale ones, simply because their individual force is small in relation to the recuperation of nature."[8] Although one wonders if he was familiar with the American automobile when he wrote those lines, the desire for small-scale, "organic" technologies shot through environmentalists who were most interested in energy issues.

A set of intertwined beliefs came to circulate among these alternative energy proponents. Technology critics like Jacques Ellul, who argued that "modern technology become a total phenomenon for civilization, the defining force of a new social order in which efficiency is no longer an option but a necessity imposed on all human activity,"[9] proved influential. Technology critic Lewis Mumford called for a new practice and theory of technology, a "biotechnics" that would emphasize "qualitative richness, amplitude, spaciousness, and freedom from quantitative pressures and crowding."[10] Among alternative energy proponents, the traditional engineering measures of efficiency and low costs were not seen as the most important requirements for technology.

Donella Meadows, the famous author of the *The Limits to Growth* report, argued for deployment near where it was used so that its users would also bear its costs. "The world works a little better any time we manage to make the invisible visible, embed real costs into prices, and impose the consequences of decision-making upon those who make the decisions," she wrote.[11] This is very nearly the opposite of what most technological companies try to do.

And when this heady mix of attitudes was turned on energy, it generated a fascination with solving the needs of heating and cooling one's shelter without recourse to centralized power systems. Electrical power could be generated right on site with wind or small-scale hydroelectric generators or, by the end of the decade, photovoltaics.

Huge nuclear reactors were the antithesis of the soft and organic and small scale. Built by huge corporations, associated with the military, and supported by an antidemocratic, militarized technocracy in the Atomic Energy Commission (and its descendents), it is not hard to see why being against nuclear was an obvious choice for appropriate technologists. "A minefield of ominous problems awaits the further expansion and international proliferation of nuclear power. These problems—environmental, social, economic, managerial, and regulatory—are at the heart of the nation's gigantic political tug of war over the future of nuclear energy," one 1980 environmental reader stated.[12]

Thus, the idea of "technological progress" came under intense scrutiny. If the narrative of American technology had been triumphant and teleological, an arrow pointing from Pilgrims to Apollo, many rejected the idea that technologies were getting "better." That had interesting consequences even in areas about which one would have thought appropriate technologists would have been remarkably enthusiastic.

Baldwin, normally pitch-perfect and succinct on why a tool or idea was good or bad, struggled to offer advice about solar photovoltaics. In 1980 he wrote, "I'd wait a year or two." Although you could get panels, they were "expensive (still) and I keep hearing sad stories of failure." However, "The price has been coming down fast and continues to fall at a rate that makes me suspect that I'd be owing more on a panel bought today than a brand new one would cost three years hence. Moreover, the photovoltaic business is fraught with breakthroughs and rumored breakthroughs." The very progress of the technology—that it was not perfect nor timeless—was an issue for Baldwin. "We aren't there yet," he almost concluded before admitting, "my own humble abode, an Airstream trailer wired 12 volts, is powered by a modest panel feeding a battery." Were we there yet or not? Baldwin can't decide, see-sawing back and forth a half dozen times in just a couple hundred words. Photovoltaics were expensive and changing quickly but they were also conceptually perfect: "simple, quiet, few building code problems, no moving parts, mounted on a wide variety of buildings, potentially mass producible in huge quantities."

The photovoltaics waffle highlighted everything difficult about the intentionally ambiguous use of the word "tool" in *Whole Earth*. If a tool was a timeless thing, a photovoltaic panel certainly did not fit the definition. If a tool was a good idea that was a means to a particular kind of lifestyle, a photovoltaic panel certainly was. "I can't recommend a brand yet, and if I did, the information would probably be obsolete by the time you read this," was Baldwin's final hedge.[13] Later, Stewart Brand tacked on an addendum recommending panels by Arco, which was owned by an oil company.

Photovoltaics were and are high-technology electronic devices. They did need to be constantly "new and improved," and they are unlikely to be repaired with vise-grip pliers or anything else that you could find in an 1850s shop. It was much easier for tool freaks to deal with wind energy, where an excellent wind machine had already been created: the Jacobs wind generator. Marcellus Jacobs had manufactured the small wind-electric generators in the 1930s and 1940s before the backcountry was electrified.[14] Enthusiasts rediscovered the machines in the 1960s and 1970s and began to speak of them as embodying gospel technological truth.[15] They were seen as the best small-scale wind generator. "At this writing, the Jacobs wind machines remain the best available despite the 40-year old design," Baldwin avowed.

THE SOLAR AGE

"There are elements of a religious revival in Appropriate Technology," wrote Witold Rybcynski, a thinker who was both part of the movement and harshly critical of it. "It is a strange mélange of Marxism, Puritanism, and something called Buddhist Economics."[16] Rybcynski highlighted that appropriate technologists supported the energy sources they did primarily because of the ideas they held about them working rather than because they worked. If solar technologies were a bit more expensive or required more time and effort, it didn't matter because they would help set society back on the right course. The occasional hassle of living in a solar heated home or fixing up an old wind generator *was part of the point.* These technologies were as much a social statement as an engineering ethos.

In the late 1960s eco-anarchist Murray Bookchin argued for a "liberatory technology." Innovation paired with ecological imagination would allow people to "regain the sense of oneness with nature that existed in humans from primordial times. Nature and the organic modes of thought it always fosters will become an integral part of human culture," he wrote. "It will reappear with a fresh spirit in man's paintings, literature, philosophy, dances, architecture, domestic furnishings, and in his very gestures and day-to-day activities."[17] It was a powerful vision for environmental types who were getting tired of apocalyptic rhetoric about pollution, pesticides, and population growth.

Certainly, there are shades of solar transcendentalism, but radical solar advocates like Steve Baer, founder of the legendary company Zomeworks, saw solar technology as a powerful force for social—not just personal—change. Baer dramatized the power of solar power to change America with a fictional vignette in his fascinating book of solar facts and stories, *Sunspots.* In this story, at a demonstration in an unnamed town a large van pulls up and begins handing out mirrors to the protesters with the invocation, "Give 'em some sunshine." The protesters then proceed to use the combined heat of eight hundred–foot square mirrors to burn a police car and start one hundred fires during the Sun Riots. "The lower offices of city hall and the police department have been gutted by fire. Black streaks surround the windows which are now shiny with aluminum foil," Baer wrote. "The police unable to confiscate mirrors; the matter is in the courts."[18]

Weaponized solar supporters breaking the government's monopoly on violence is merely the most extreme of the visions harbored by appropriate technologists. "They thought that a decentralized society, relying on solar power, would be a more just society, promoting self-reliance, and reducing the income gap," historian Frank Laird has shown.[19] Many were determinists, figuring that technologies change people and not the other way around. As the historian Kirk summarized their thinking, "Technology used amorally and unecologicaly created the social and environmental problems of industrial capitalism; therefore technology used morally and ecologically could create a revolution toward a utopian future."[20]

Perhaps this belief blinded appropriate technologists to the actual desires of the people who might buy their products. "What extremes of temperature within a house are comfortable?" Steve Baer asked. "In a dry climate like Albuquerque I believe year lows and highs of 55 degrees and 85 degrees are perfectly easy to live with inside a house—especially if you have warm spots such as fireplaces or stoves to stand next to when it is chilly."[21] That most people do not want to live in homes that get that hot or cold and are willing to make generous use of the HVAC industry to avoid doing so should come as no surprise.

Political scientist Langdon Winner indicted appropriate technologists for their failure "to face squarely the facts of organized social and political power. Fascinated by dreams of a spontaneous grassroots revolution, they avoided any deep-seeking analysis of the institutions that control the direction of technological and economic development."[22] This means that while they were out picking new lifestyles as a deep revolutionary gesture, the rest of the world carried on passing laws and building new power plants and developing new stuff like computers, fiber optics, and the sports utility vehicle. The appropriate technology movement wasn't squashed; it was simply sidestepped or ignored. Although Baer's company has succeeded in carving out a niche and continues to this day, few others did.

One honest alternative technologist, Peter Harper, well-known in the UK and the States for his work in the field and writing in the magazine *Undercurrents*, blasted the idea that everyone would want to live in solar homes powered by meager amounts of wind and run on large amounts of self-provided labor. Harper wrote that

> it's my guess that, faced with the real choices and real costs, nearly everybody would opt to stay in the straight society as things are at the moment. Apparent successes of cheap alternative technologies have on the whole been achieved through hidden subsidies of time or resources which could not be generalized throughout society. At the moment, only those with very unusual tastes (such as for Spartan living), or those who place an extremely high value on environmental purity, or those that think that the relative positions of 'straight' and 'alternative' economics will change markedly, would find it rational to pay the full cost of ATs.[23]

Few have been willing to make those tradeoffs throughout the years. It's not that there isn't a thriving off-grid energy community or that the neo-homestead movement has gone away; rather, it's become a niche, like the group of people who love Italianate homes and Porsche 911s. Industrialized societies allow both types of people to happily coexist and subsidize them both in different ways. Harper went on to challenge,

> Why not have computers, power stations, TV, hi-fi sets, and laboursaving devices? What exactly is the case against them? Why not distribute electricity through a grid? When were you last oppressed by the local electricity board? How much of what you need can you get in a community of 10? 100? 1000? 10,000? Make a list of all the things you have in your home. How many could you do without? How many could you make yourself? How many are made on a massive scale and would cost five times more made in any other way?[24]

Bespoke technologies have a difficult time competing in a world that expects very high levels of service and reliable products from companies. The products alternative technologists built rarely made it

through the "various stages of development, debugging, and deployment that any new technology needs."[25] Basically, many didn't work. In
1980 a surprisingly hopeful book called *Solar Failure* detailed the numerous ways that systems broke. The authors' intent was not to "dis-
courage the layman by bombarding him with an endless list of failures"
but rather to help out in "truly triumphant" solar undertakings that
would avoid the mistakes that had been made during the 1970s. Looking back, the catalog of errors and problems go a long way toward explaining why the solar age remained more slogan than reality.
"Performance experiences have identified major problem areas resulting from: 1. Unfamiliarity of designers with requirements of the solar
energy system. 2. Unexpected mechanical failures (leaks, corrosion, ab-
normal wear, etc.) 3. Unexpected structural failures (collector storage,
interface with the building structure, etc," the authors wrote.[26]

Further problems occurred with collectors, tanks, valves, pipes,
pumps, flow meters, heat exchangers, and "systems installed without
thought given to fluid expansion" or improper air pressure, and a dozen
other things.

What's more, a lot of the problems did "not become apparent until
the system" had been operating for a while. As a consumer, it was clearly
a bad situation, as Baldwin recognized in 1980 in noting that the new solar industry was "a strange mixture of dedicated ineptitude, cool profes-
sionalism, and fraud. Right now, it's pretty hard to sort it all out—so
many of the companies involved are small and local that investigating
each one is virtually impossible."[27] If the spiritual leaders of the movement could not navigate the waters, what was anyone else going to do?

Solar technologies were not the only products that suffered from
these problems. In the concluding story in their book, *Design for a Limited Planet*, which sold 100,000 copies in the mid 1970s, Jon Naar and
Norma Skurka relate the story of the installation of a Jacobs wind machine atop a dilapidated tenement on New York's Lower East Side. The
triumphant story, filled with photos of happy people of color, instructs
that the building "has become a model for other homesteading efforts
in poverty areas, particularly where landlords abandon buildings because of escalating maintenance costs."[28] The tenants, along with massive infusions of activists, successfully fought Consolidated Edison for
the right to sell their excess electricity supply back to the grid. Their

case even attracted the attention of Ted Kennedy in Congress, who called the turbine atop the building, "the little windmill that could." Wind historian Robert Righter declared it "a symbol people need not be entirely reliant upon centralized utility monopolies."[29]

However, according to a 2008 *New York Times* article that looked back on the building's legacy, the windmill never worked very well. "Either wind speeds were too low to generate sufficient power or turbulence from gusts produced a deafening noise from the windmill and caused the building to shake," we learn. During a blackout, the windmill couldn't even provide light for common areas. After about ten years of this off-and-on service, it broke in what the *Times* describes as a hurricane.[30]

This type of thing led to some harsh condemnations. "To the pragmatic observer . . . who is interested in solving problems, AT has little to offer," Rybcynski wrote in 1978. "It is a movement that is long on polemics and pitifully short on actual accomplishments."[31] The lesson that many learned in the late '70s was that environmentalists were a lot better at stopping stuff from getting built than building it themselves.[32]

But appropriate technology's legacy is not as negative as some would suggest. They laid the groundwork for many of the eco-businesses that rose during the 1990s and provided a valuable training ground for many of today's senior green-tech contributors. The very idea of *beating* the bad technologies with good stuff was, in itself, powerful. "These designers wanted to fight fire with fire," Kirk has concluded. "They wanted to resist technocracy and frightening nuclear and military technology by placing the power of small-scale, easily understood, appropriate technology in the hands of anyone willing to listen." In lining out such a program, they made a valiant effort toward "reenvisioning environmental activism from the ground up."[33]

If they failed to create a true alternative society, their ideas still spread into existing institutional structures, informing the passive solar work of Art Rosenfeld and the Lawrence Berkeley Laboratory as well as the government of Jerry Brown. Ty Cashman, a member of the New Alchemists, a prominent appropriate technology group, was the one who got the renewable energy tax credits passed that created the market opportunity for both the infant wind industry and Luz's solar farms, as discussed in chapters 14 and 24.

chapter 20

Technology

THE TRUTH IS THAT the most important technological work done to capture solar energy during the 1970s didn't come from the tool freaks. Instead, an oil company funded the research that enabled solar photovoltaic cells to get radically cheaper. Esso, which re-branded as Exxon in 1973, supported the work of Elliot Burman, a chemist who led the Solar Power Corporation's efforts to cut the costs of solar energy. Under Burman's leadership, the company drove down the cost of producing a solar module about an order of magnitude, from $100 per watt in 1970 to $10 per watt in 1973.[1] The Solar Power Corporation successfully adapted what had become a space-bound technology for regular commercial operations on earth.

The Exxon team reexamined the entire solar module for cost efficiencies. First, they looked at the biggest material cost—the silicon—and switched to cheaper wafers rejected by the exploding semiconductor industry. Then they eliminated unnecessary steps in the processing of the silicon to waste less of the expensive stuff. Finally, they rethought the module encasement. For terrestrial applications, the modules just didn't need to be as rugged, so cheaper materials could be used.

The development of Solar Power's first module is a prototypical development venture. The key cost problems were identified and solved within the constraints imposed by the physics of the materials and the logic of the marketplace. Burman did exactly what many of the solar home builders and appropriate technologists could not: organize and execute an ambitious technological program to make solar power radically cheaper.

In so doing, Exxon brought the cost of solar down enough to find an actual marketplace in far-flung locations like oil rigs. It may not have been a large market, but it was big enough to keep companies interested in competing and developing the technologies further. The advances that came in the years following these big cost reductions proved to be of a different kind than what Burman was able to do so quickly.[2]

University of California Berkeley energy researcher Gregory Nemet found that the two largest factors in cost cutting between 1975 and 2001 were increases in the efficiency of modules in converting sunlight into electricity and capturing the economies of scale that come with building bigger plants. While scientific research may have continued to improve efficiencies without a market for photovoltaics, the scaling effects were made possible by the deep price drops brought about by the Solar Power Corporation and a select few other companies.

This is the promise of technological development. Advances lead to price drops, which encourage scaling, which leads to more price drops, and so on and so forth. "The cost of photovoltaics has declined by a factor of nearly 100 since the 1950s, more than any other energy technology during that period," Nemet noted.

As we saw in the last chapter, even the most technology-friendly of the solar advocates of the era struggled with how to think about photovoltaics. Amory Lovins and the *Whole Earth* editors reported fairly wild-eyed projections about the possible arrival of "cheap photovoltaic cells," but neither emphasizes them as an energy pathway. "This development may be imminent and should be closely watched," Lovins wrote in *Soft Energy Paths*, "though the analysis in this book nowhere assumes this or any other solar-electric technology." Much more important, Lovins thought, was the reduction of electricity use in general.[3] As noted, *Whole Earth's* James T. Baldwin waffled on whether or not his readers should purchase photovoltaics, but he also ran a 1981 review of the book *Photovoltaics* that predicted "photovoltaic systems installed on the roofs of residences in the U.S. will be fully economic—delivering electricity that costs five to ten cents per kilowatt hour—by 1986 without tax rebates." This turned out to be massively optimistic.[4] Appropriate technologists largely treated photovoltaics as a possible *deus ex machina*: great if it happens, but incidental and outside their own work.

Their attitude, shared with the broader environmental movement, toward other big-budget solar research was even more negative. Ray Reece's book *The Sun Betrayed: A Report on the Corporate Seizure of U.S. Solar Energy Development* is a monument to the feeling that government money was going to the wrong projects and that the Exxons of the world were doing more to stifle the development of solar energy than to help it.

Certainly, the government's record of supporting small enterprises with research funds was deplorable during the time when lots of money was available for solar research. The vast majority of funds disbursed by the Department of Energy and its predecessor, the Energy Research and Development Agency, went to large corporations like Lockheed and Exxon.

Reece records a Denver scientist, Jerry Plunkett, telling a Senate committee in 1975 that "we don't have to have college professors tell us what the intensity of the sun is or that solar energy is a workable system. . . . There are workable systems on people's homes and no need for gathering research data."[5] The problems with solar energy, Plunkett said, required business innovators, not researchers.

It would be abundantly clear by the early 1980s that much, much more research was actually necessary to create wind and solar energy systems that were reliable, but few solar supporters wanted to believe that. Reece himself, following the Austin, Texas, solar pioneers he knew, was furious about the way R&D funding was allocated. The social revolution that some imagined would naturally result from solar technology itself was not happening. For them, the big companies who moved in and bought up the small solar companies were at fault. They reduced solar from a liberatory technology to another product line.

In describing the reasons why the Department of Energy gave the grants to the big companies, Reece actually made the case for exactly why big businesses were needed. They might not have been the perfect vehicles for funding, but they had the resources to pursue the kind of grinding industrial research that had made all kinds of technologies viable through the years. Reece thundered that

[corporations'] advantage over smaller "competitors" is obvious: tremendous capital resources and control over raw materials,

established factories and dealer networks, brand-name identification, unlimited advertising budgets, and high-level access to
government R&D funds. That should be quite sufficient to assure
the giants whatever degree of hegemony over the new solar market they desire. Indeed, most of those very corporate attributes
are cited by government planners as justification for transferring
the nation's solar R&D program to the large corporations, the
presumption being only they, in concert with the utilities, have
the wherewithal to execute the "commercial development" of solar energy in the United States.[6]

And Reece was not alone in his anger about the "corporate elite."
Steve Baer often launched vehement attacks on corporations' technological approaches. Baer described his experience with the International Solar Energy Society in a *Mother Earth News* article in 1976. He
wrote,

Everybody there would be talking about sophisticated collectors
and tracking systems and very exotic and expensive surfaces
that were marginally more efficient absorbers of the sun's rays
and multi-million-dollar research projects. And, usually, the
guys doing all the talking didn't have a working prototype of
anything they were spouting off about.[7]

We can only imagine what Baer would have said about a solar
company owned by Exxon, but we know what Reece said: "By what
miracle had the Exxons and Lockheeds of the land become the fount
of innovative alternative energy technologies for the federal government? What were the corporations producing in the area that was so
superior to the concepts developed by the small R&D groups whom I
had observed?

The decidedly anti-establishment stance of many solar proponents
prevented them from seeing that they would have to use the corporations to achieve a broad range of good, less expensive solar products.
After all, the alternative technology groups of the day produced nothing comparable to the large-scale technological project that the improvement of photovoltaic cells has been.

In silicon, researchers developed all kinds of manufacturing tricks and research knowledge about the material itself. In what are known as "thin-film" solar cells, which use far less material than silicon, researchers identified and began to experiment with the main materials that are now considered commercially interesting today: cadmium telluride, copper indium gallium selenide, amorphous silicon, and so forth. "The [cost] that has been reached also required a progression of substantial and creative R&D improvement in materials, devices, fabrication, characterization, and processing, leading to better device performance and reliability, and lowered systems costs," wrote Larry Kazmerski in 2006, a photovoltaics researcher at the National Renewable Energy Laboratory for thirty years.[8]

Changes of this magnitude would not have been possible without long-term research programs executed by high-level scientists and engineers, exactly the sorts of people who made the counterculture solar advocates cringe. Nonetheless, the work of those scientist-stiffs has now changed the economics that underlie all kinds of solar advocacy.

The point, perhaps, is that the desire to use renewable energy, even in the 1970s, could be separated out from the counterculture. Although we might remember solar as being the province of hippies, it was far more widespread than that, and many more types of people made contributions to what is now almost ubiquitously referred to as the "clean energy future."

Take, for example, the husband-and-wife astronomer team Aden and Marjorie Meinel. They traveled the nation promoting their idea of a massive thousand-gigawatt, five thousand–square mile National Solar Power Facility. They hoped that the plants would be finished by 2076, at which point the facility would be producing the entire nation's electricity needs at a constant five cents a kilowatt hour, or about what people paid for power in most parts of the country. It was an attention-grabbing scheme. "A pollution-free energy system that could supply the whole nation with all the power it needs forever could be set up in the Arizona-California desert," one newspaper wrote in December 1970.[9]

The Meinels were advocating what might be called a "hard solar" energy path—and they meant it to be matched to the scale of the U.S. economy. Where the appropriate technologists and solar transcendentalists tried to sever as many ties to the industrial world as they could, the Meinels' key insight was that solar could scale up faster if it found the right conduits into the existing technological systems that distributed power. They imagined a solar thermal power system much like the ones built by Luz International in the 1980s (see chapter 14) and many more companies in recent years. Parabolic mirrors would focus the sun's rays on thin, liquid-filled tubes suspended at just the right spot. Each tube would be coated with selectively absorbent materials, which would trap heat inside. The heated liquid would then be used to generate steam, which could turn a turbine just like in a traditional fossil fuel or nuclear power plant. "This isn't a new electricity system," Marjorie Meinel stressed to the Associated Press in 1972. "It's a new fuel source for generators that already exist."[10]

The scale of the project would have required more than three million acres of land. Huge aqueducts would have had to bring water from the Gulf of California and Pacific Ocean. If all had gone according to plan, that water would have been desalinated and used to "make the deserts bloom"—the surefire mark of a true American mega-scale engineering scheme.[11]

The Meinels took their show on the road, putting the full weight of their scientific credibility behind the project. Aden Meinel was the head of the Optical Sciences Center at the University of Arizona and Marjorie had a master's in astronomy and had edited scientific journals at Cal-Tech.[12] The prominence of their affiliations meant that they got audiences you might not expect, like some important officials of energy- and sun-rich regions like Saudi Arabia.

Furthermore, the Meinels didn't look down on small-scale solar. In fact, it had inspired them at the Applied Solar Energy Society conference in 1956. At that event, the giants of mid-century solar research all turned out. The University of Wisconsin's Farrington Daniels, a former nuclear scientist who had been a member of the Manhattan Project, showed off his solar cooker. The University of Colorado's George Lof promoted his ideas for solar housing. Dozens of other researchers and groups came together under the hot Phoenix sun as the Meinels wandered the exhibits.

However, one thing stuck with them about the approach of the Applied Solar Energy Society, interesting as they found their colleagues' work. "All of the solar was for individuals. You could have it for your house. You could have a solar cooker," Aden Meinel said. "We thought that solar should be large-scale instead if it was going to have any national impact."[13]

But large-scale solar advocates came under heavy attack. An article in *Science* noted that federal solar research had supported "large central stations to produce solar electricity in some distant future" while largely ignoring smaller technologies. "The massive engineering projects designed by aerospace companies which dominate much of the program seem to have in mind the existing utility industry—rather than individuals or communities—as the ultimate consumer of solar energy equipment," the authors wrote.[14] They suggested that the government's program was struggling with the nature of solar energy, which was different. "Solar energy is democratic. It falls on everyone and can be put to use by individuals and small groups of people."[15]

Although solar energy does fall everywhere, it doesn't fall equally everywhere. Nor are the dynamics of the energy industry erased by its availability: The utility business was and is a major consumer of solar energy equipment. The criticisms also assumed that large-scale deployments of solar collectors and other energy sources would be occurring quickly, disregarding any criticism of solar's technological readiness as bias or pessimism. The experience of green technologists over the past thirty years suggests that distributed solar and wind power *were not* ready for industrial-scale production during the mid 1970s. More and better research might have changed that, but without massive technological progress, they would not have made meaningful contributions to the nation's electricity system.

Even simple things like passive solar housing design needed technical and analytical infrastructure as provided by Art Rosenfeld and the physicists (see chapter 17). Large-scale solar farms are on the verge of a massive commercialization all over the world, primarily with expertise gained at the Luz plants (see chapter 14), which themselves drew on federal R&D into solar thermal technologies (see chapter 13). The Meinels themselves could not secure research funds to extend their work beyond the laboratory scale, and eventually they gave up on solar energy, but not before laying out the gameplan for later solar farms.

The innovation ecosystem of the 1970s wasn't perfect. The idea of funding small, high-tech companies instead of either large, high-tech companies or low-tech solar hopefuls did not take hold. What is un-questionable, though, is that technological progress during the 1970s set up the later successes of green technology and made renewable energy a natural place to turn when energy prices and global warming concern began to rise in the early twenty-first century. We can also say that the psychological and even spiritual territory explored by the solar tran-scendentalists and tool freaks was important, too. They helped change conceptions of technology as having to be big and technocratic. Many of the themes that they explored, like decentralization and personal empowerment through technology, crop up again in the personal com-puter revolution, which historian Fred Turner brilliantly shows in his book *From Counterculture to Cyberculture* to have grown, in part, on the substrate of the counterculture.[16] More directly, many countercul-tural energy types grew up and became policy wonks. And their funda-mentally different interpretation of the value of fossil fuels and solar energy has informed the political left ever since.

A solar society might not have been realized, but a base of knowl-edge and a set of institutions were created through which the dream of a more perfect power would rise again.

V.

Innovation and the Future

chapter 21

Google's RE < C Challenge

COAL IS A FLAMMABLE ROCK. You can put it in a pail and carry it to your house. Or you can pack it into one hundred cargo containers and load it onto a train. It's plentiful and spread pretty evenly around the world. Take a pound of it. Burn it completely and you would generate ten thousand British thermal units (BTUs) of heat. Dirty and heavy, coal has exactly one thing going for it: price. "The problem however is that for much of the world, coal remains both a cheap and portable (and storable) source of energy," energy analyst Gregor Macdonald wrote in 2009. "In fact, coal has a nasty habit of pricing itself *just below* nearly all other energy sources. This is why I've called coal a kind of anti-hero."[1]

But there are reasons coal is cheap. Economists call them "externalities." What this means is that a business charges some of its cost to society without having to pay for it. A classic example played out in Chicago and Pittsburgh around the turn of the century. People with coal boilers had a choice: They could burn the soft, smoky bituminous coal or they could pay a little more and burn anthracite coal, which burned more cleanly. If you were a manufacturer trying to maximize your own profits, you picked the soft stuff and let all that soot go up the smoke stack. There were some mitigation technologies available, but they cost money, so you didn't use them. In a "free" market, you could do what you wanted. But what if your next-door neighbors happened to make or sell textiles? As you maximized your profit burning your cheaper coal, the soot and particulates would pour out onto their linens, ruining their products.

This is a classic example of the problem of pollution for a capitalist society and it was not a hypothetical scenario. City residents around the world organized against the use of soft coal, passing laws and prodding smoke inspectors to do *something* about the soot belching onto them. They argued that the societal costs of the soft coal exceeded any money that individual soft coal users saved. Chicago's smoke inspector, F. U. Adams wrote in 1894 that

> they insist that its consumption entails an annual damage greater than the difference in cost between soft and hard coal; they declare that the smoke nuisance is a positive menace to the health of citizens, that it has resulted in an alarming increase in throat, lung, and eye diseases. They point to ruined carpets, paintings, fabrics, the soot-besmeared facades of buildings, and to a smoke-beclouded sky, and demand that the Smoke Inspector do his plain duty under the law.[2]

The point is that manufacturers had every economic incentive to burn the cheaper stuff and it was cheaper precisely because society had to bear some of the cost, a phenomenon that Garrett Hardin first brought to widespread notice in his short 1968 essay, "The Tragedy of the Commons." "The rational man finds that his share of the cost of the wastes he discharges into the commons is less than the cost of purifying his wastes before releasing them," Hardin wrote. "Since this is true for everyone, we are locked into a system of 'fouling our own nest' so long as we behave only as independent, rational, free-enterprises."[3]

This particular aspect of coal use has continued to wind through American history. In the past the best solutions have been a combination of technical advances and new regulations. For example, long-distance electrical transmission allowed the locus of coal burning to move outside central cities. Although this helped the immediate environment and health of urban areas, the shift to out-of-town power production meant that the impacts that did occur became less visible. Smoke was still coming out, but people no longer saw it soiling *their* linens.

Eventually, emissions caught up with this fix. Sulfur and nitrogen oxide that coal plants pumped out form nitric and sulfuric acids in the atmosphere. Then they fall back to earth with rather nasty consequences for plants, animals, and people alike as acid rain. In the latter half of the twentieth century acid rain began to damage human and natural environments at the continental level. It didn't matter how far away they put the plants.[4] But again, technology and regulation came together to reduce coal's impacts. Federal legislation in 1990 created a system of tradeable allowances that commoditized emissions. With this price on sulfur emissions, "scrubbers" were installed on manufacturing plants and the problem was mostly fixed. When the cost of the sulfur started to hit companies' bottom lines instead of America's, utilities figured out how to cut the emissions, even though it's taken decades.[5]

For a long time polluters were willing and able to believe that "the solution to pollution is dilution."[6] Their reasoning was essentially that the earth was big, humans were small, and the globe could take it. Though some scientists tried to call attention to the potential of the carbon dioxide produced in combustion to derange the atmosphere, few industrial types believed that humans could cause the greenhouse effect until well into the 1980s.[7] (And of course many still don't.)

Nonetheless, the mounting evidence of climate change—and its biological impacts—shows that humans truly have reached the global scale with our pollution. Dilution is no longer possible: The nasty effects of coal burning have hit their geographical limit. The world is warming and weirding as the physics of the new earth change longtime patterns in how water and wind flow across our globe.

Unlike previous episodes with coal, there is no natural governmental body to regulate its consumption. Cities can't do it. Countries can't do it. The United Nations has been unable to do it. Worse yet, the more that developed countries do to destroy demand for coal, the cheaper they actually make it, encouraging greater consumption in countries that have more pressing concerns like providing basic lighting services to their citizens than mitigating climate change.[8]

The solution, which Google formalized into a neat equation in late 2007, is pretty simple.

RE < C

Renewable energy (RE) has to become cheaper than coal (C). In Europe and even the United States, it may be possible to make the C more expensive. But in the Indonesias, Chinas, and Indias of the world, the Russias and Ukraines, whether or not that will be possible remains unclear. At the very least the political negotiations are going to be tough. And in the end, what's necessary is to actually keep the coal in the ground, not reduce the rate at which it's burned. This is not going to be easy. "It's hard to win a fight against a cheap BTU," Macdonald likes to say.[9]

The RE has to get a lot cheaper. Much of this book is dedicated to the process of how technologies get better: We make more of something and each one gets cheaper; we figure out how to optimize our factories and supply chains so we lower our costs; financial institutions see our power plants as low risk and the cost of the financing we need to build them comes down. Many researchers have tried to quantify all of these factors under what they call a learning curve. Such curves quantify how fast a technology gets better and/or cheaper. The problem is that they are notoriously inaccurate and filled with discontinuities: One year it looks like a technology is getting cheaper quickly, but then it slows down the next—and vice versa. Few people claim to be good at predicting when those inflection points might occur on technologies evolving over the course of decades.[10]

Making RE cheaper than coal is going to be difficult for a few structural reasons. Take a look at a rushing river. Let's take the Columbia, the fiercest of all the rivers in North America. The river exists because the sun evaporates water from the world's oceans, which is carried in the form of clouds into the continental United States. The clouds cool and when they hit the Rockies, they are forced to rise and get even colder. Precipitation falls and the water seeks the path of least resistance, running on a meandering path from the mountains south and west through the West. Through numerous combinations, the Columbia's path was created, wearing a bed into the crust of the earth over thousands of years. The force delivered by the water derives from gravity, which pulls the water down to sea.

Humans used to know the river by the work that it took them to fight it or the work it let them escape when they worked with its force. The same could be said of the wind of the plains or the sun of the deserts. In his book *The Organic Machine*, Richard White wrote,

Engineers can measure the potential energy and the kinetic energy of the Columbia with some precision, but early voyagers . . . recognized the power—the energy—by more immediate if cruder measures. They measured it by the damage it did as it threw ships or boats or bodies against rocks or sandbars. And they measured it by the work they had to perform to counter the river's work.[11]

The river did not respect human boundaries of force. People drowned. People died. The water of the river, with no anthropomorphizing necessary, played by its own rules. The complex thermodynamics of the water flow was beyond human understanding.

A wild river served humans only incidentally. We had to find ways to satisfy our own needs within its physical behavior. Boat captains learned how to use the natural flows and eddies to do what they needed to do, but they were ultimately riding at the pleasure of the river. Mark Twain called piloting a "wonderful science" that let him read the wild Mississippi like a book.

Even if we dam a river, some part of it remains outside human control. The places where we burn coal are some of the most highly engineered places on earth. They use special types of steel to isolate the coal burning from the outside world. They are completely manmade systems.

Green-tech machines, however, are harder to build and even test. As far back as the 1830s, engineers at the Franklin Institute in Pennsylvania were trying to build a better water wheel. They created an elaborate machine to run tests and model how the water's force worked with different water wheels. A chamber for water was connected to a series of valves, a bell, and a timing device. The input water was standardized so that the same amount fell on each wheel. Measuring how much water the wheel could raise how far, a common measure of power, similarly quantified the output of the wheel.

The experimenters ran 1,381 experiments with the machine, and the publication of the results was considered a tremendous success. The problem was that the human parts of the system weren't really the variables that mattered. Rainfall, the particulars of a river's location, seasonality—those were the things that made a big difference. The machine was just one small part of the overall operation. Too many variables could not be accounted for. The same goes double for solar power and triple for wind. The overall energy system has to be shaped so it can accommodate natural energy flows.[12]

Green-tech plants require tight integration with a place in the world. To effectively use them, we have to scientifically understand those places. Progress on that score has been slow and steady. The first tool to even measure the amount of solar energy hitting the earth at a particular location, the bolometer, wasn't even developed until 1878, eight years after coal power overtook water as the dominant manufacturing power source.

Much research has been done since then, but better data will go a long way toward dealing with the problems of renewable energy. The best-case scenario for our relationship to wind and solar energy can be summed up by Samuel L. Jackson's comments in the movie *Jackie Brown* about his rather unfaithful part-time girlfriend: "You can't trust Melanie, but you can trust Melanie to be Melanie." That is to say, she's not dependable in the absolute sense, but because he knows how she works, he can dependably predict that behavior in general, regardless of what it might be in particular. Likewise, the more we learn about how natural energy sources work, the more we can learn to trust them to be themselves, even if they are less faithful than coal.

If renewable energy systems aren't perfect, they do have one major thing going for them: They don't use fire. The material we put into them is not destroyed when we use it. No waste is generated, and no carbon dioxide is released into the atmosphere as they operate. The same cannot be said for coal, natural gas, and oil.

In the purple prose of the 1930s writer Paul Lewis, waterpower (and we could extend the same idea to wind and solar) allows humans to get power "right down here on the rock-bottom of cosmological causation." There's no need for "the explosive combinations of oil or gas which these fuels enable him to make use of. [Man] is down on the everlasting elements—water and magnetism." If this is a harnessing of nature, it's certainly a gentler version than that wrought by Prometheus. "The conquering of Nature!—" Lewis concluded, "but into her friendly bosom again even as we conquer her."[13]

In more scholarly terms, White, author of *The Organic Machine*, suggests that humans have never actually conquered nature. We have merely become more and more deeply enmeshed in the world. "Labor, rather than 'conquering' nature, involves human beings with the world so thoroughly that they can never be disentangled."[14]

Consequently, renewable energy sources entangle us ever more with the world in which we live, and that's not a bad thing. But making RE beat C around the world is going to take a tremendous amount of innovation. This section looks at ideas, technologies, policies, and methods that could help the world get there.

Society's intricate mechanisms share a symbiotic relationship with energy. We live in a time when their billions of people are colliding with new ideas, technologies, and the few limits that matter. The continuing information revolution, breakdown of traditional economic models, and decreasing availability of oil in an urbanizing world are deep trends. If the past is any record, changes this big will push civilization to a new equilibrium point. Different ways of doing things will make sense, and the world will change. How it does so, however, is up to us.

chapter 22

The First Megawatt and Failing Smart

I N THE WEE HOURS of March 26, 1945, the wind was blowing at a sleepy five miles per hour, far too slow for the turbine to make electricity. Harold Perry, a construction foreman, had been working nonstop for the twenty-three grueling days since the wind power plant had gone back online after some repairs. That night, an elevator carried Perry one hundred feet up through the oil derrick-like tower to the small, armored building that housed the controls for the world's largest wind machine.

In the years before World War II, this machine, the Smith-Putnam wind turbine, stood as a testament to the power of human—and American—ingenuity. A decade before, the Soviet Union had built the world's largest wind turbine, a one hundred–kilowatt machine, and now, just a decade later, the Yanks had constructed their own version ten times more powerful. "Vermont's mountain winds were harnessed last week to generate electricity for its homes and factories," read the September 8, 1941 issue of *Time*. "Slowly, like the movements of an awakening giant, two stainless-steel vanes—the size and shape of a bomber's wings—began to rotate."[1]

Over the next two years the turbine ran through hundreds of hours of testing, often pumping power onto the electrical grid. The project's engineers were sure that, technically, the machine worked. Unluckily, a bearing broke and the war prevented its replacement until 1945. By spring of that year, with the war waning, it was back up and running.

When running at maximum speed, the turbine could produce 1,600 horsepower—that's 1,250 kilowatts. It would take fifteen thousand human beings working hard to generate that much energy. Just imagine them spread out in the blue-green valley beneath Perry, pulling on ropes or pushing rocks up the hill, a platoon of Sisyphuses—or, more realistically, slaves. Harnessing nature has its benefits.[2]

Understanding how ridiculously grand the project really was is important. Its scale—ten times as powerful as the very largest turbine and a thousand times more powerful than most of them—was almost unimaginable. To plan an equivalently ambitious project now would mean setting out to build a machine that pumped out sixty-five megawatts. This was a small group of inventors attempting to make a leap into a different future through breakthrough technology.

And the strange thing is that they succeeded.

Time concluded its article on the project with a hopeful half-prediction, "New England ranges may someday rival Holland as a land of windmills." This was, after all, merely the prototype for whole lines of turbines.

Perry's job was to watch over the turbine and make sure that everything ran smoothly. The turbine had built-in methods for "coning" out of the wind to keep it from spinning too quickly, but it seemed like a good idea to have someone around . . . just in case. During the day, he could stand behind the rotating blades in a flannel shirt and a hardhat, staring out at the unspoiled expanses of rural New England. Old films show the blades—the bomber wing look-alikes—beating a rhythmic, majestic *whomp-whomp* right in front of his face. But it was dark just then—three in the morning. He would not have seen much out there.[3]

Atop the rural Vermont mountaintop known as Grandpa's Knob, aloft in that tower with wings, Perry didn't know that the grandest wind experiment in the first few millennia of human existence was about to fail. For ten more minutes, he would be sitting atop the world's most famous evidence of renewable energy's bright future.

The unprecedented project was built up from nothing, practically conjured by Palmer Putnam, an MIT-trained geologist with no formal education or experience in wind power. He was a fascinating character, a clean energy entrepreneur seventy years ahead of his time. Vannevar Bush, Franklin Delano Roosevelt's science adviser, had high praise for this engineer-of-all-trades, calling Putnam a "go-getter" in his autobiography and noting that he "had some of the characteristics of the best type of promoter in industry. He was well liked by men with lots of drive and often disliked by those with less."[4] His friends called him Put, after the Greatest Generation traditions of the day.

Before his project, windmills had just pumped water for farmers in the boonies or charged the batteries of rural radios so they could pick up the AM stations that brought news across the lonely, whistling prairies. The people who sold windmills marketed to ranchers and farmers; their advertisements appeared in magazines like *American Thresherman and Farm Power, Agricultural Technology*, and *Successful Farming*. They were symbols of autonomy from the centralized systems of the electrical industry. They were simple and Western and rugged.[5]

But that's not the kind of turbine that Putnam had in mind. After looking into the designs of the past, he immediately decided that the economics of scale dictated that he build a wind turbine with seventy-five-foot blades, the largest in the world. It would generate more than a megawatt of power and feed it onto the grid, working in tandem with hydroelectric plants to smooth out the intermittency of the wind and the seasonality of water generation.[6] No one had ever accomplished either of the latter tasks, and most people working in the wind industry at the time were probably too sane to try.

Putnam was no fool, but he had been trained to think of himself as a player on the world stage. His father, George Putnam, ran the most venerable publishing company in Manhattan and was a celebrated Civil War vet. A giant in his time, the elder Putnam is remembered most for writing *The Little Gingerbread Man*. His mother was the first dean of Barnard and cofounded the New School in New York. His cousin was an Arctic explorer who married the aviatrix Amelia Earhart. They were society folk: Between Palmer's birth in 1900 and his father's death in 1930 the *New York Times* wrote dozens of articles about their fortunes. His people were deeply enmeshed in the power structures of the time, mingling with presidents, generals, authors, intellectuals, and moneymen.

But Palmer Putnam wasn't a swell or a powerbroker. He was a tinkerer and maker in the long tradition of such self-identified types in the New England region. His only foray into the glamorous New York world of publishing ended in ruin.

When not-quite-as-rich-as-he-seemed George died, he left his only son $90,000 dollars, equivalent to perhaps a million or two today.[7] The thirty-year-old Palmer, a would-be geologist, used his cash to buy out a cousin's stake in the business, thereby becoming president of G. P. Putnam and Sons. He arranged a merger with an upstart publisher, but with the Depression sweeping the country, it didn't help the family business. They tanked. In 1934 he filed for bankruptcy in New York, claiming debts of $75,000 and no assets.[8]

Before the ill-fated corporate endeavor, he had been a hardworking but carefree MIT master's student, content to learn the brown magic of engineering and geology. He had even written a master's thesis with the rakish title, "A Reconnaissance Among Some Volcanoes of Central America."[9] After his corporate failure, he returned to the familiar space of engineering. In the year of his bankruptcy was when he conceived of his wind machine—and began making plans to build it.[10]

Short on actual capital, his seersucker pockets were overflowing with cultural capital. Putnam loved a good sail and the good company that came with it. And it turns out that this bit of cultural knowledge might have meant more to the project than any technical knowledge he gained at MIT.

Many men might have dreamed of a huge wind machine, but Putnam was the only one who had access to the boardrooms and argot of the favored classes. Once he had hit upon his bankruptcy-induced idea that "a windmill to generate alternating current might reduce the power bill," he was off to the races, calling on upper-crust contacts across the Northeast, asking them to trust in the power of the wind based on their memory of it in their sails.

By 1938 he was able to sell a General Electric vice president, Thomas Knight, on the project, because, as Putnam wrote, Knight "had sailed Down-east all his life and was immediately attracted by a proposal to use recent developments in aerodynamics and other fields to harness the wind on a large scale."[11]

Knight dispatched one of his men, Alan Goodwin, to convince the head of the New England Public Service Corporation, Walter Wyman, to sign on. Goodwin's sales pitch to "the one man in New England who had both the authority and the vision to push such a project" was indirect, almost Scooter Libby–like, in its evocation of nature. Putnam wrote in his book, *Power from the Wind*, that Goodwin made this appeal: "Mr. Wyman, just look at the way the wind is blowing those trees.

There is a lot of force there and it is all going to waste. Man has used wind for centuries to blow himself around the oceans but he has never harnessed it for power on a large scale."[12]

Goodwin's point is well noted. Two modern economic historians have estimated that wind power, mostly tapped by boats, was the dominant energy source for most of the nineteenth century. In 1850, by their figuring, windpower did more work helping distributors drag goods around the States than waterwheels and coal combined.[13] Wyman was swayed. He set up Putnam with a subordinate company, the Vermont Power Service—and launched the project into the big leagues.

With someone to buy the power they hoped to generate, the project soon found a financial backer—the S. Morgan Smith Company. Wyman happened to know that with the nation's rivers damn near dammed up, the hydroelectric turbine company was looking to expand into a fresh market. They already knew how to do water—why not add the air, too? When all was said and done, the largish family business headquartered in York, Pennsylvania, spent $1.25 million, or upward of $15 million in today's money.[14]

With financial backing and Bush's imprimatur, Putnam started to work the MIT network to find star engineers. He had been introduced to E. N. Fales, who was one of the first people to start thinking about how to apply the knowledge that engineers had gained from prop planes to the windmill. Fales's key advance was to replace the multi-bladed mill with two simple, propeller-style blades. That let the blades spin faster—six to ten times faster—which created way more power. By the early '20s Fales was already writing that his design enabled "competition with gasoline farm-lighting plants."[15]

These advances in aerodynamic knowledge convinced Putnam that if he just took Fales's design, made it big, and hooked it to the grid, he could generate electricity as cheaply as coal or hydropower. He was apparently quite persuasive. His next hire was John Wilbur, head of civil engineering at MIT, to serve as the project's chief engineer.

"MEAGER AND UNCERTAIN" DATA

The construction of such a novel machine was lousy with difficulties, some internal but many external. The exigencies of a country preparing for war caused delays and trouble, but the desire to finish and monetize the turbine ahead of the fast-approaching war appears to have accelerated the pace of R&D beyond what prudence would have dictated.

Evaluating the performance of designs was also painstakingly slow because the engineers of the time lacked the computers to do tough math quickly. Cal-Tech aeronautical engineer Homer J. Stewart was assigned to calculate the effectiveness of different rotor designs. In a 1982 interview he recalled the computational process. "I'd compute the designs, step-by-step numerical integration for hours on end. It was the sort of thing that my pocket computer can now do," Stewart said. "It's a fairly messy problem; on the pocket computer it takes an hour's time to compute one power output at a given wind speed for a given design to a reasonable precision. It took weeks in those days."[16] Today, a similar problem could be completed nearly instantly on a MacBook, allowing engineers to optimize the design by running millions of calculations.

Their effort also suffered from gaping data holes. Gathering engineering-grade environmental data is just flat-out difficult—a lot harder than the task for coal plant engineers, who get to design a structure and measure inside *that*. Wind engineers have to deal with natural conditions. Their work is *in situ*: It's out in the wild. Now we have hundreds of weather stations to record precise information about the wind—and historical data stretches back decades. Wind maps show the average movement of air at excellent resolution. New sensors and computer models have transformed the field of aerodynamics, too. And even with all this, getting just the right placement for wind turbines and farms is still difficult.

Back then, the information available to Putnam was "meager and uncertain." People knew that some places were windy, but that was about it. To actually design and engineer a turbine, they needed to know not only how strong the wind was but also how consistently it blew. That data didn't exist. The wind, though we normally think of it more like the lines behind a running cartoon, is three-dimensional. It blows in gusts and at odd angles and with plenty of chop. Engineers call this "the structure of the wind," and Putnam's team—or engineers thirty years later, for that matter—didn't really have a good understanding of how it worked.

Even just figuring out which mountain to stick the turbine on was tough. In their initial analysis, Putnam recruited a biologist to look at how bent the trees were! From that, they tried to back into the average power of the wind at different points scattered across New England. They could erect anemometers to measure the wind speed for a short period of time, but even that precaution was mostly disregarded. In the end, Grandpa's Knob was selected merely because they had to get moving on the project and it seemed like a pretty good spot. They hadn't even had a chance to build a cute little plaster model of the location and stick it in a research wind tunnel like they had done for other locations. Years later, analyses from the 1980s suggest that understanding the flow of the wind at Grandpa's Knob would have been incredibly difficult given the state of meteorological science but that the site was, in retrospect, a poor choice.[17]

A few weeks after they had picked the site, they finally installed an anemometer. It showed the wind's power to be a mere 10 to 30 percent of what they had predicted, but (luckily) bad data from a rogue observer on another mountain convinced the team that the entire region was just experiencing a freak low-wind season. "It is quite likely that we have this observer to thank for the Smith-Putnam Wind-Turbine experiment," Putnam dryly commented. "If it had been known that not only was there no anomaly, but also little wind at those elevations below which we did not fear ice, it is likely that the experiment would have been abandoned out of hand."[18]

But they didn't know any of that, so the team soldiered on.

The crowd of MIT and Cal-Tech engineers were forced to order major pieces of the turbine in May 1940 before they had finalized the basic design of the machine. For example, they had to order the parts of the machine that would hold the blade to the rotor without knowing how heavy the blades were going to be. Putnam contended that "this calculated risk turned out badly and contributed to the later failure."

In late 1940 parts began to rumble in from all the old glorious industrial towns of the country. The tower and support structure came from a bridge builder in Ambridge, Pennsylvania. Another company built the blades in Philadelphia, and a third put almost all of it together in Cleveland and shipped it by rail to Rutland, Vermont, a few hours northwest of Boston.

To build a huge machine, one needs huge pieces. Getting the gargantuan items the ten miles from the town to the foot of Grandpa's Knob required some slick engineering and a lot of goodwill from local authorities. The steel structures were heavier and wider than the local roads could accommodate, so they pulled out the power lines on the sides of the asphalt and temporarily reinforced the bridges they traveled over. The transport took ten nerve-wracking trips, but within a few days they had miraculously assembled all the components at the base of the mountain.

The two thousand–foot trek to the top of the mountain proved more difficult. There was no road winding up the mountain, so they built one themselves. The self-made route was treacherous and steep, approaching a 15 percent grade in some spots. The team pushed tractor-trailers loaded with parts from behind with a bulldozer and pulled from the front with a half-track, a sort of demi-tank with standard wheels up front and treads in the back.

On the way up, the construction project experienced its first major setback. At one curve, a forty-three-ton girder tipped off a trailer and fell into a ditch by the road. It took three weeks just to set up the rigging to get the hunk of metal back onto the trailer. Eventually, though, through weeks of brutal effort, all that shaped steel, which had originated across the Northeast, was on the top of the mountain.[19]

The final stage of construction—connecting the blades to the massive structure and wiring everything up—went smoothly. On August 29, 1941, the blades of the world's first megawatt wind turbine spun for the first time. A year and a half before, it had been just an idea in Putnam's head.

THE FREE ENTERPRISERS

If the feat was improbable, that was actually part of the point. The group of men (yes, they were all men) that assembled around the project was united by their love of invention. They wanted to prove that they could harness nature—for profit. They were, in their own words, Yankee free enterprisers.

The people of New England, the Yankees, had long thought of themselves as the world's preeminent innovators. Their brains were the intellectual engines of the American Industrial Revolution. They were Silicon Valley–style technologists, entrepreneurs, and financiers when Silicon Valley was just a few Catholic Missions somewhere near the physical feature known as the San Francisco Bay. "The wind turbine is notable as the physical result of a project conceived and carried through by free enterprisers," Vannevar Bush wrote, "who were willing to accept the risks involved in exploring the frontiers of knowledge, in the hope of ultimate financial gain."[20]

Given the bewildering technological changes that occurred from 1900 to 1945, summoning the sense that there was progress in the world was not difficult. A similar sentiment seized California in the wake of the silicon revolution: Don't resist the country's energy hunger, but feed it. Things are getting better.

Mark Twain's Connecticut Yankee is a model of the free enterpriser breed, wandering King Arthur's Court, bragging that he could make "all sorts of labor-saving machinery. Why, I could make anything a body wanted—anything in the world, it didn't make any difference what; and if there wasn't any quick new-fangled way to make a thing, I could invent one—and do it as easy as rolling off a log."[21]

New Englanders, you might say, felt a bit of pride in this ability to remake the natural world into "labor-saving machinery" with the ease that mere mortals rolled off logs. Putnam embodied the Yankee to sell his project and he got it built.

But what turned out to be more important was the way Putnam failed.

The broken bearing, which was routine and not a failure of design, shut down the turbine from 1943 to 1945. For two long years the project sat while the world fought, quite possibly hurting the durability of the materials. During that time, Wilbur had calculated that the bracing where the blades attached to the electricity-generating apparatus was under tremendous stress. He was worried about its structural integrity.[22]

So, maybe, as Perry sat atop the structure, he was a little worried about the Smith-Putnam. Perhaps he pulled out the Sunday paper to take his mind off the noise of the machinery and the possibility that something might go wrong when he was alone with the wind machine.

World War II dominated the headlines. The war, at least in Europe, was ending and everyone knew it—Patton had smashed across the Rhine with ease—but what kind of peace would result was a mystery. The Bomb remained undropped; atoms had not been used for war yet, let alone for peace.

On the home front the social disruptions of the world's biggest catastrophe were beginning to subside. Men returned home. The standard order snapped back into place, beating back the inversions and deviations of wartime. Plans were laid for postwar abundance, but the actual shape that the future would take—fossil fueled, energy intensive, suburbanized, car driven—had not yet been determined.

A country away in San Francisco, Nancy Pelosi turned five years old.

The next decade would see American society transformed. A boom—more or less mirrored across the industrialized world—in babies, income, chemicals, and comfort would lead to the greatest run-up in energy usage the world has ever seen.

But in 1945 war-time habits were still in play. The deprivation it brought to industrialized countries forced them to use energy and materials with a care that they have never approached again. And still the country was 500,000 tons short of coal, prompting confrontations between coal mine operators and workers about how to pick up the slack. There was a need, or at least room, for alternative energy sources.

Wind, it turned out, would not be one of those sources for another forty years.

At exactly 3:10 a.m. on March 26, 1945, after more than 1,100 hours of operation, the Smith-Putnam turbine experienced an epic failure. One of the turbine's blades broke clean off and went sailing 750 feet through the night. The force of the breaking blade threw Perry off his feet as the unbalanced machine shook like the bridge of the *Star Trek Enterprise* when it was under attack. Putnam dramatized the scene:

Suddenly he found himself on his face on the floor, jammed against one wall of the control room. He got to his knees and was straightening up to start for the control panel, when he was again thrown to the floor. He collected himself, got off the floor, hurled his solid 225 pounds over the rotating 24-inch main shaft, reached the controls, and brought the unit to a full stop in about 10 seconds by rapidly feathering what was found to be the remaining blade of the turbine.[23]

A photo taken the next day shows the enormous blade on the ground, men walking and crawling near it like the Lilliputians around Gulliver. The caption reads simply, "The Blade That Failed."

THE BLADE THAT FAILED
AND THE PROJECT THAT DIDN'T

Seeing catastrophic failure in the Smith-Putnam turbine is easy. What went wrong is as obvious as a seventy-five-foot blade lying on the ground. The turbine's record, however, was a mixed bag for renewable energy advocates in political debate. On the one hand, the existence of the turbine made wind power seem more real as a large-scale source of electrical power than it ever had been before. On the other hand, at the 1951 Congressional hearings to provide increased wind power funding, one historian notes, some "legislators considered Putnam's blade failure to have proved the whole endeavor a washout."[24] The machine's failure played right into the hands of those committed to other forms of electrical production—fossil, atomic, or solar. Putnam himself later advocated the use of atomic and solar power to replace fossil fuels in the long run, devoting only a few dismissive sentences to the potential of wind power in a sweeping energy analysis he wrote for the Atomic Energy Commission in the early 1950s.

But for the wind engineers who discovered Putnam in later years, the turbine wasn't a failure. "Interest in developing large wind-electric generating systems in the United States was stimulated primarily by one man, Palmer C. Putnam," a crisis-induced 1974 NASA research report on alternative energy found. Even though his turbine didn't ultimately succeed, he left a remarkable base that was really the only thing that the inventors of the '60s and '70s had to build on.[25]

Putnam failed well because he created data on which inventors could build the future. In a startlingly progressive move, the company that bankrolled the project assigned their patents to the public domain and asked Putnam to write a book detailing what happened so that others could continue the work. They made the wind data they had gathered from the region public. This turns out to have been immensely helpful for later generations.

Without the unique experiment, nothing would have been known about large-scale systems. Less data equals more risk—and risk is expensive in big power plant projects. By gathering data on what did and didn't work, Putnam saved enormous amounts of time and money for subsequent researchers. Now an entire industry is standing partially on Putnam's shoulders: In 2008 wind employed more people than coal mining.[26]

Today, the rough economic times have led to a shakeout in green industry. The winners are being separated from the losers. In some cases, this is because the technologies aren't working or aren't scaling up. But the losses are not always engineering issues. Business problems, exacerbated by the financial crisis, are foreclosing technological possibilities before they have a chance to play out. A bad economy combined with a glut of companies devastated the green-tech industry in the mid-80s. As we'll see in later chapters, whole types of engineering knowledge and data were lost. Recent green-tech supporters have had to reinvent a lot of wheels.

Given the likelihood that the vast majority of today's green-tech companies will fail, even if some wildly succeed, that data sharing remains rare is troubling. What will happen to the data from failed wind and solar companies? Unlike the Smith-Putnam turbine, they might take the key to the next breakthrough to their corporate graves. Gregory Nemet, innovation researcher at the University of Wisconsin, argues that we need a "countercyclical" program in the United States to capture knowledge during busts to save for the booms.[27] Furthermore, DOE funding could come with some information-preservation strings attached.

Putnam, for his part, returned to wind energy at the end of his life. He moved to Atascadero, California, a few hours south of San Francisco, and attended DOE conferences on wind power, where he spoke to the newcomers in the field. With the wind industry's success, Putnam's initial hopes for his own turbine, inscribed in the headline of an article in *Power* magazine—"Wind-Turbine Power Plant Will Be Rebuilt"—came true: Tens of thousands of large wind turbines have been built since his pioneering work, even though nearly all of them followed a slightly different design.

His own turbine was never rebuilt, nor any more on its exact model. Within six months of the catastrophic blade failure, the S. Morgan Smith Company shut down its wind program. They had run out of money for investigating wind turbines with no guarantee of a return. They pulled the plug instead of plunking down the $300,000 that Putnam needed to build a new prototype.

The blade was carted off, the turbine torn down. A cell phone tower now adorns Grandpa's Knob.

But the foundation of the great wind machine remains.

chapter 23

What Green Tech Can Learn from Nuclear Power's Rise and Fall

O
N JUNE 10, 1964, Lyndon B. Johnson rode through the streets of Worcester, Massachusetts, cheered by 175,000 well-wishers as he was on his way to give a commencement speech at Holy Cross. Looking out over the football stadium's cheering masses, dressed in the traditional scholar's robe, the Texan lawyer delivered a paean to science and technology's power to transform the lot of the world's poor for the better.[1]

Spattered with bits of Christianity, he identified three "ominous obstacles to man's effort to build a great world society—a place where every man can find a life free from hunger and disease—a life offering the chance to seek spiritual fulfillment unhampered by the degradation of bodily misery."[2]

While paying lip service to disease, he concentrated on two other problems for which he had the same solution: poverty and "diminishing natural resources." The way forward against both these menaces was nuclear power. Johnson stated,

> There is no simple solution to these problems. In the past there would have been no solution at all. Today, the constantly unfolding conquests of science give man the power over his world and nature which brings the prospect of success within the purview of hope. To commemorate the United Nations 20th birthday, 1965 has been designated International Cooperation Year. I propose to dedicate this year to finding new techniques

for making man's knowledge serve man's welfare. Let this be the year of science. Let it be a turning point in the struggle—not of man against man, but of man against nature.[3]

There would be a technological fix for the world's problems. There could be prosperity for all through exploiting nature more intelligently, largely through "our new capability to use the power of the atom to meet human needs." He declared,

It appears that the long promised day of economical nuclear power is close at hand. In the past several months we have achieved an economic breakthrough in the use of larger-scale reactors for commercial power. And as a result of this rapid progress we are years ahead of our planned progress. This new technology, now being applied in the United States, will be available to the world.

Through the magic black box of science, nuclear energy would be transformed into American soft power throughout the world. Johnson concluded his thoughts on nuclear energy by stating,

The development of the large-scale reactor offers a dramatic prospect of transforming sea water into water suitable for human consumption and industrial use. Large-scale nuclear reactors and desalting plants offer, in combination, economical electric power and useable water in areas of need. We are engaged in research and development to transform this scientists' concept into reality.[4]

With unlimited power and water, all the world could be a Monticello—open for life, liberty, and the pursuit of happiness. Wealth would not have to be redistributed because there would be enough for everyone to live an American lifestyle even if, as Johnson noted, that would require producing natural resources at one hundred times their production levels.

As in the original Dwight D. Eisenhower "Atoms for Peace" speech, the specter of nuclear destruction—which, like it or not, was an

American invention—was redeemed by the utopian visions of a perfect power. "We now can join knowledge to faith and science to belief to realize in our time the ancient hope of a world which is a fit home for all," Johnson concluded. "The New Testament enjoins us to 'Go ye therefore and teach all nations.'"[5]

Thus, nuclear power, long-supported by the American government with subsidies, was officially enshrined as the American energy technology of the future. The reactor was a cheap, clean, necessary answer to the problem of the bomb and the opportunity of the future.

Or so Johnson's story went. It was a grand American narrative: Science! Technology! Progress! Economic growth! Unlimited everything! What's not to love? It's more than a bit like the one we are telling ourselves about green technology.

Unfortunately, the kernel on which it was built—the "economic breakthrough" of nuclear power—was more truthy than true. That did not make it any less effective. Nuclear power had been in the offing for more than a decade, but in the five years following his speech was when it became a reality. Nuclear power generates about 20 percent of America's electricity, greatly reducing the carbon intensity of the energy system.[6]

Beginning in the early 1970s environmentalists tended to look at the negative aspects of this rise in power.[7] They criticized the subsidies that the industry received or its close connections with the military. They questioned the costs of building new nuclear reactors. They called out nuclear engineers and scientists for bias against solar energy. They brought up the specter of meltdowns and nuclear proliferation.[8] Likewise, many nuclear advocates like Edward Teller or Atomic Energy Commissioners like Gorman Smith pooh-poohed the efforts of solar pioneers. Solar energy was too diffuse, too land intensive, and not suitable as the basis for an advanced civilization.[9]

Alvin Weinberg, former head of Oak Ridge National Laboratory and longtime deep thinker about energy, once wrote that the ultimate energy battle is between uranium and the sun.[10] When all the fossil fuels are gone or otherwise unavailable, those are the two energy sources that will be left to support whatever humans are around. And during the 1970s the more extreme supporters of the two technological paths took to the ramparts a little too early.

There is, however, an alternate history of energy thinkers who have sought diversification of the country's energy system—and particularly the reduction of coal use, which is the dirtiest fuel we have. Farrington Daniels, perhaps the most influential solar technologist of the mid-century, also holds a patent for a nuclear reactor and worked on the Manhattan Project.[11] James Lovelock, made world famous by his Gaia theory of earth, has long been a nuclear supporter.[12] Art Rosenfeld, the "godfather" of energy efficiency, supports nuclear power as long as it's cheaper than moderating demand.[13] Plenty of nuclear advocates shared the attitude that both solar and nuclear power could coexist in a future energy system. Weinberg himself asked the leading question, "Is it not the most sensible course to aim for a system that depends on some combination of solar and nuclear?"[14]

These solar scientists recognized that although huge nuclear power plant designers and rooftop solar installers might have vastly different ideals and business models, they did have a common enemy: coal. Plenty of nuclear advocates pondered the same question that Weinberg asked, and many saw their power as "clean" in comparison with coal long before carbon dioxide was a primary concern. Coal, after all, has been shown to be far more destructive to the environment than nuclear power, even before climate change became a primary environmental issue.[15]

What's important here, though, is not the relationship between solar and nuclear power, which is a story that has been told many times over. The success of nuclear power is at least as interesting as its comparative problems or links with military operations. After all, this technology that was not economically competitive with coal became a major contributor to the American energy system in only twenty years. But nuclear power didn't succeed only because of its usefulness to the military's bomb-making complex, nor did it succeed because, as Johnson's speech suggests, nuclear power got cheap quickly. The strategies its promoters employed and the structural factors underlying its emergence provide a fascinating model for green-tech advocates.

What we'll focus on here is how nuclear power transformed from a sluggish technology looking for a market into a huge industry from 1965 to 1975. Just like solar and wind advocates now, nukes, as they called themselves, battled coal interests. However, the difference for nuclear power pushers is that they won, thereby garnering tremendous governmental and industrial support.

Before the Joint Committee on Atomic Energy in 1965, Joseph Moody, president of the National Coal Policy Conference, told the assembled Congressmen that his testimony was "one of the most ineffective things I do." They laughed in his face.[16]

This, despite the fact that economic historian Steven Mark Cohn concluded in his study of the rise of nuclear power that there was no monetary basis for the size and scope of the American government's support. "The bottom line is that there was no compelling economic case for large public or private investments in commercial nuclear power plants in the 1950s and 1960s," Cohn wrote.[17] In 1959, for example, a private Atomic Energy Commission (AEC) briefing for President Dwight Eisenhower concluded, "For our own economy, with but few exceptions, we do not need atomic energy in the foreseeable future."[18]

IF YOU BUY IT, IT'S CHEAP

The *New York Times* ran a story about Johnson's speech on page one under the headline, "Johnson Reports a 'Breakthrough' in Atomic Power." They followed up with a series of stories, as did the other major newspapers.[19] Word of a breakthrough in the cost of nuclear power was big news because everyone had been waiting for economically feasible nuclear power for a decade. After the heavy promotion of the early nuclear power days—exemplified by Walt Disney's classic nuclear cartoon, *Our Friend the Atom*—nuclear power had stalled out with just a few demonstration plants in operation. The coal lobby, however, smelled blood. In March of 1964 the coal industry assailed nuclear power, saying Congress needed to remove "the sheltering umbrella of Government subsidies."[20]

General Electric and Westinghouse, who had helped build America's military and civilian nuclear program, were getting antsy that their knowledge would go to waste. "Our people understood this was a game of massive stakes, and that if we didn't force the utility to put those stations on line, we'd end up with nothing," as John Gitterick, a GE vice president, later told *Fortune*.[21]

Out of this corporate desire to capture rents on a technology that only a few companies could provide came the "economic break-through." As soon as the words left Johnson's mouth, scientists at national laboratories around the country knew what he was talking about, even though he was a few months late with the announcement.

When a *Chicago Tribune* reporter called Stephen Lawrowski, associate director of Argonne National Laboratory, the scientist told him that the president must have been talking about the guaranteed price that General Electric had offered Jersey Central Light and Power for the Oyster Creek plant. That announcement had "caused a flurry" in scientific circles because the price GE was charging for the plant—$68 million for the 515-megawatt plant—made the plant economically competitive with fossil fuels.[22] Yet the scientists knew from the available evidence that nuclear power was far from economically competitive in mid-1964.[23]

However, instead of setting the *Tribune* reporter straight, Lawrowski simply punted, saying "The New Jersey plant is a significant milestone in nuclear power progress because it has affected thinking not only in America but also in Europe."[24]

The price was a door-buster, a loss-leader, an advertisement for a nuclear age that had not actually yet arrived. The so-called "turnkey" plants, as they later became known, probably cost Westinghouse and General Electric over $1 billion, though they did not say that at the time.[25]

Coal officials told the *Wall Street Journal* that GE had "priced the Oyster Creek plant at less than cost." A GE executive denied that, claiming the company would "make a slight profit unless we run into some unforeseen difficulties."[26] British and Russian engineers also called the estimates into question—and French officials unsuccessfully tried to get details out of GE. But American news accounts, though they reported those foreign doubts, always made sure to note the bias that national competition could introduce into other countries' expert opinion.[27]

None questioned the U.S. expert corps' own Cold War sympathies. Yet in retrospect, no less a nuclear proponent than Alvin Weinberg saw that the scientists and engineers had taken leave of their senses. Against a backdrop in which Lyndon Johnson's most effective campaign ad against Barry Goldwater featured a child picking flower petals until there's a harsh cut to a mushroom cloud, the nuclear community was caught between visions of apocalypse and utopia. As Weinberg recalled in his sharp memoir:

> I find it hard to convey to the reader the extraordinary psychological impact the G.E. economic breakthrough had on us. We had created this new source of energy, this horrible weapon: we had hoped that it would become a boon, not a burden. But economical power—something that would vindicate our hopes—this had seemed unlikely. . . . Because we all wanted to believe that our bomb-tainted technology really provided humankind with practical, cheap, and inexhaustible energy we were more than willing to take the G.E. price list at face value.[28]

Newspaper reporters, with the help of sources within the nuclear industries, came up with stories to explain how prices could have fallen so far, so fast. But like a trend piece about raising chickens in Manhattan, they were little more than anecdotes strung together by plausibility and the public's desire to believe.[29] Although they reported doubts about the breakthrough, they were often run deep inside the paper whereas the optimistic pieces led the sections of the paper. Even the most skeptical piece, a September 1964 article by *Washington Post* reporter Howard Simons, noting that "not all experts accept General Electric's figures," only questioned the figures within 12 percent.[30] In reality, nuclear power would end up costing not $104 or $1,040 per kilowatt of capacity but more than $3,750 per kilowatt by the mid-1980s.[31] Perhaps Lewis Strauss, then-chairman of the AEC, overstated the case when he told a crowd of science writers in 1954 that "Our children will enjoy in their homes electrical energy too cheap to meter," but his optimism was obviously widely shared within the nuclear establishment.

The country's political leaders were more than willing to believe and promote these technical promises. It was a wonderfully convenient solution to an America battling Communist agitation across the world. After all, could the Russians offer cheap nuclear power that turned the atom into electricity and oceans into fresh water?

The foreign policy positives of civilian nuclear power had been deeply embedded in American rhetoric since the dawn of the Cold War. When Dwight D. Eisenhower delivered his "Atoms for Peace" speech in 1954 to the UN General Assembly, he spent the majority of his talk on the subject of destruction before he turned to the title of the talk and the potential of creation. Eisenhower stated,

> The United States knows that peaceful power from atomic energy is no dream of the future. That capability, already proved, is here—now—today. Who can doubt, if the entire body of the world's scientists and engineers had adequate amounts of fissionable material with which to test and develop their ideas, that this capability would rapidly be transformed into universal, efficient, and economic usage.[32]

This was pure faith in technological promise, but hey, it was the half-century of the automobile, airplane, telephone, nuclear bomb, and widespread electricity. Plus, the Soviets were hot and heavy to develop nuclear reactors themselves. Eisenhower thought history was on his side and that the future demanded his particular idea of progress.

ESTABLISHING FUTURE FACTS

That's because nuclear proponents convinced politicians that their set of future facts about energy in America were correct. They said energy usage would soar and they had nice graphs to back it up. Their vision was expansive, expensive, and rather brilliant. Technical reports came out purporting to show energy "needs" for Americans in the future that were spectacularly high. From the early 1950s until the energy crises of the 1970s, politicians accepted as gospel truth nuclear proponents' overblown visions of America's energy needs emanating from the nation's national laboratories and the AEC. Legislators continually delivered high-levels of steady funding to nuclear research.

Of course, the political relationship ran both ways. The AEC knew what the government needed and the government knew what the AEC needed. In both cases, the answer was: Don't stop believing!

In 1960 the AEC, which had as its mandate to promote the commercialization of nuclear power, projected that Americans would use 170 quadrillion BTUs in 2000.[33] In reality, that year Americans used about 99 million quads of energy. And we still do. Imagine adding 70 percent more power plants, cars, and buildings to our current energy infrastructure. It's nearly unthinkable. After reviewing energy demand growth forecasts made by the AEC later in the decade, Glenn Seaborg, the commission's head, made the consequences of such inexorable growth clear. "Nuclear power has arrived on the scene, historically speaking, in the nick of time," he argued, because of the "projected demand for power based on population growth and increasing per capita consumption of electricity."[34]

Nuclear power was the only solution to "the energy problem" as it was conceived. "I do not think, therefore, that anyone can seriously believe we could rely on coal as our major source of power as we enter the twenty-first century or that we should not develop with all due urgency the best systems for producing nuclear power," Seaborg wrote in *Peaceful Uses of Nuclear Energy*.[35] If coal, the most plentiful fossil fuel source we use, wasn't up to the long-term task of meeting our energy needs, who could argue that energy conservation or wind or natural gas had any chance of working?

Dozens of technical reports by energy analysts affirmed that the country's energy demand was growing at prodigious rates, and they projected that growth to continue far into the future. This was not disingenuous. People who studied energy in the middle of the century had grown up in a century of nearly straight-line increases in energy consumption. The problem was, however, that they just compounded the growth of energy demand year after year after year. If we project anything in that manner, the numbers get really big, really fast. The top-down analysis was convincing even though it ended up very, very wrong.[36] They ultimately whiffed by wide margins on what the future energy system would look like, but not before fears that only nuclear could maintain our society created a clear path for lawmakers.

Despite the occasional call for the free market to work, the opposite happened. For example, nuclear power plant operators are indemnified by the U.S. government for catastrophic disasters (the Price-Anderson Act), thereby lowering their insurance rates. They were given preferential access to markets for borrowing money. There was plenty of informal and regulatory help to go with the R&D and commercialization boosts.[37] In effect, the government socially engineered the cost structure of the industry so nuclear could compete with coal, which got to dump all its extra costs, such as air and water pollution, into the environment.

But even then, convincing utilities that they needed to go nuclear wasn't easy until General Electric hit on the genius idea of guaranteeing a fixed price to risk-averse utilities, effectively subsidizing the cost of the construction. And Oyster Creek was born. If they could just build a ton of plants, they could learn and scale and standardize: Costs would drop. Westinghouse matched GE's pricing, and what came to be known as the "turnkey" plants were built. In the bandwagon market that followed until 1973, utilities ordered more than two hundred nuclear reactors. Nuclear power had arrived.[38]

But the turnkey plant prices did not reflect the actual costs of building a nuclear power plant. As the years wore on, that nuclear power was not as cheap as coal and other fossil fuels became increasingly clear: The prestige of the nuclear authorities began to fall; nuclear whistleblowers came forward; environmental risks were reassessed, perhaps too stringently; the protest movements of the 1960s turned their attention to nuclear power and all the centralization of power it represented.[39] It turned out that Americans were ready to extend democracy to technocratic decision making, and they did not like what they saw from the nuclear industry.

The nuclear industry operated as a closed network of thinkers and analysts, disregarding outside critiques of their methodologies and not taking the serious issues of nuclear power seriously enough. "One result of the regulators' professional identification with the owners and operators of the plants in the battles over nuclear energy was a tendency to try to control information to disadvantage the anti-nuclear side," a former AEC commissioner admitted in the early 1990s.[40] The very agency charged with regulating the industry—the Atomic Energy Commission—was also charged with promoting it, and that's just the most obvious conflict of interest. Nearly everyone involved in assuring the public of the economics, safety, and environmental wisdom of atomic power was also involved in promoting atomic power. Not all of them had economic interests at stake, but few were disinterested observers.

The coalition of scientists, reactor builders, and utilities neglected the social aspects of their technology. A more subtle type of blindness to the effects of actual success afflicted the nuclear crew as well: Success surprised them.[41] Though they believed in the engineering idea of scale—bigger is better, bigger is more efficient, bigger is cheaper—with unerring faith, they tended to ignore the problems that scale would bring. The complexity of local and global politics, safety, construction, waste management, and plant siting were all underestimated. And that all cost money. The cost of the plants rose for many reasons, not just those that pro- or anti-partisans like to highlight.[42]

Learning to run the plants well also took a long time. The capacity factors of those huge nuclear plants—how often the plants were actually generating electricity—were shockingly low. They hovered in the 58 percent range, which means that if we visited a plant on ten random days, it would not have been running for four of them. Since then the capacity has improved and is now over 90 percent, which is a testament to how good technology can become over time.

But that came too late. The energy futurology that served the industry so well began to break down. Energy demand growth did not just continue accelerating as they had anticipated. All the projections from 1960 through 1980 fell short by an average of 40 percent. We didn't need as much energy as we had anticipated. As a result, the vision of the nation's future that sold nuclear power never panned out.[43]

Higher-than-expected costs, worse-than-expected operation, the meltdown at Three Mile Island, and the Chernobyl disaster all obviously hurt the industry with the public. A less well-known event might have occurred on October 5, 1983, when Cincinnati G&E announced that its Zimmer nuclear station would need 2.8 to 3.5 billion more dollars and two to three years of further construction time. Previously, the utility had claimed the reactor was 97 percent complete. "That news was the first of many disastrous nuclear crises that followed," wrote Leonard Hyman, an investment banker who worked with the utility industry. "Utilities tottered on the brink of bankruptcy, scrambling for funds to complete troubled projects, or to salvage what they could from huge investments in projects that had to be cancelled despite the billions that had been sunk in them."[44]

Investors got the message: Nuclear power was not a good investment, so they scurried away. The First Nuclear Era, as Weinberg called it, was over. No new reactors would be built in the United States for more than twenty-five years.

LEARNABLE LESSONS

Green-tech advocates can and should learn from both the rise and the fall of the nuclear industry. It's the last time a major new energy source came online, and its successes and mistakes were epic. Several lessons emerge from this history.

First, the short-run economics of a technology are not always what drives governments to support them or businesses to invest in them. We bank on technological learning and skill, and we often have goals beyond the immediate economics of a technology that could be attained by a different energy source. For example, the foreign policy objectives of the United States helped nuclear power gain prominence and support, despite its expense. Beyond the military R&D that went into reactors, being able to offer a world torn between capitalism and communism a way to abundance was important. Nuclear power became a major part of selling the American dream.

In green technology, advocates argue that incentives make sense because when technologies get deployed faster than the market might otherwise pull them in, the cost to produce them falls faster, too. Let the industry fatten on government dollars and the technology will get better. Renewable energy will play an important part in American foreign policy, too, as the world's governments have broadly agreed on the need to limit carbon emissions going forward.[45]

Third, the quantitative and qualitative projections about what the world's future energy system will look like were and will be the key to decision making. These projections are presented with the air of facts. If you can get your facts installed early, it's mighty difficult to beat back those assertions. Nuclear proponents were excellent at getting their visions of a very high-energy world treated as an assumption. With current energy assumptions, the projections of both the Department of Energy's Energy Information Administration and the International Energy Agency set the playing field for the future. Now, however, there is a new key factor: the projections of global warming consequences produced by the Intergovernmental Panel on Climate Change.

Maintaining popular support for a technology as it grows through (very public) growing pains is difficult, particularly if hopes about the technology have been raised.

When Barack Obama stood before MIT in autumn of 2009 to deliver a major speech on energy, he echoed Lyndon Johnson at Holy Cross: "Countries on every corner of this Earth now recognize that energy supplies are growing scarcer, energy demands are growing larger, and rising energy use imperils the planet we will leave to future generations," Obama said. The answer to these problems lay in innovation in clean energy. He went on to state that "the nation that harnessed electricity and the energy contained in the atom, that developed the steamboat and the modern solar cell" will "lead the clean energy economy of tomorrow."[46]

Again, the problem was energy, the environment, and a rising population. Again, the solution was technology. Only now, it's the wind and the sun and energy efficiency that are going to fix the world's problems. The promise is the same—can the end result be different?

Climate legislation, which would increase the cost of fossil-fuel generation in one way or another, is the obvious next step to "make renewable energy the profitable kind of energy in America." Whereas incentives for nuclear power were about decreasing the risks and costs associated with it, green technologists are using disincentives for other types of energy production combined with federal R&D programs that are expected to push green-tech costs down.

Green-tech advocates, spreading out from the venture capital centers of the country, have succeeded in installing green technology as the carrier of technological and economic progress. As General Electric and Westinghouse provided key expertise and financial support for an immature technology, so now do venture capital firms and newly created energy companies. It's a different, far less centralized model, but the effect is the same: Momentum has swung to the side of green advocates in the public eye. Those same companies are also pumping billions of dollars into developing technologies that for decades languished in labs or were cast off as side projects.

The road to generating 20 percent of the country's electricity, though, is going to require hundreds of billions of dollars more. Costs will have to be driven down and green jobs delivered to those who have been promised them. Local environmental concerns over the use of the Mojave Desert or Cape Cod or West Virginia hilltops for wind farms threaten to derail the coalition of people supporting green technology.

There will be problems with the technologies, and if history is any guide, their adoption will almost certainly go slower than the optimists expect.[47] In a democracy like the United States, with its messy, human processes, perhaps Alvin Weinberg provides the best advice for would-be green-tech revolutionaries at the end of his sad elegy for the nuclear industry of his day. "It never occurred to us that we not only had to engineer a nuclear system that was sound technically, but also one that was 'right' politically," he concluded.[48]

chapter 24

The Five-Cent Turbine and the
Siren Call of the Breakthrough

THE 33M-VS was not just another wind turbine; it was the answer
to the problem of fossil fuels and it was the would-be foundation
of the next great American technology company.

Designed between 1989 and 1993 by Kenetech, the world's largest
wind turbine maker, along with a consortium of electric utilities, it was
touted as a technological breakthrough. The "marriage of aerospace
and microelectronics" would finally make the electricity produced by
capturing the flow of air across the earth as cheap as that produced by
burning rocks, liquids, or gas.[1]

On paper, the 33M-VS was capable of making electrical power for
five cents per kilowatt-hour. Five cents wasn't just cheap: It was fossil-
fuel cheap. The company promoted the new innovation as the Five-Cent
Turbine and treated it as a technological savior for the problems beset-
ting the wind industry in particular and renewable energy more
broadly.[2] "For years, the wind industry's goal has been to produce power
at rates similar to oil's: roughly a nickel for a kilowatt. Machines now operating in California can produce energy at 7 cents per kW," *Time* summarized. "In areas of consistent high winds, the next generation,
currently being deployed, will bring that cost down to 5 cents by 1995."[3]

The 33M-VS was the leading light of that next generation. With a
name that evokes a spy plane or a sports car, the turbine had cost $70
million to develop at a time when being a millionaire still meant something. It was bigger than its predecessors and produced three times
more power than Kenetech's previous generation.

Aside from the size, it didn't look much different from what most people know as a modern wind turbine: white and oblong with three thin blades facing into the wind. What mattered was the insides. Thanks to a completely redesigned electrical system, the 33M-VS was going to use the wind's gustiness, not fight it. In handing Kenetech their environmental innovation award for 1993, *Discover* magazine rhapsodized,

> Unlike previous turbines, the 33M-VS is rigged to roll with the wind's punches. When gusts whip the rotor, the generator shaft is free to speed up in response. As the shaft's rotation speed changes with the wind, the alternating current that flows from the generator swings up and down in frequency. But between the generator and the utility grid lies an electronic power converter. This device first converts the variable-frequency current to direct current, then switches it back to alternating current at a fixed 60 cycles per second. So the generator feeds an even current to the utility grid. And the wind gust problems—wear and tear and wasted energy—have all but blown away.[4]

During the first wind-power boom, engineers discovered all sorts of things about how strange, powerful, and three-dimensional the wind is. In some of those early places like the Altamont Pass, the movement of air from the cool Pacific into the warm San Joaquin Valley shredded the first machines put in its path.

The worst part was that in those extra-strong gusts of winds was where the most electricity was to be had. The power in the wind varies with the cube of its speed, so if we power ten lightbulbs at one speed, we can power eighty at double that speed.[5] Makers of electricity-producing turbines had long been hamstrung, though. They couldn't use the upper range, or the lower for that matter. They had to design for the middling wind because electricity is generated by the smooth, steady rotation of special magnets. Give those magnets the jitters and the power quality degrades. Given that wind power already had a spotty reputation with the utilities, turbine makers had to sacrifice production to ensure that they delivered the steady electricity required by the grid.

So no matter how fast the wind blew on almost all the turbines designed in the 1980s, the rotor, which is the part of the turbine that transmits mechanical energy to the generator, had to rotate at the same speed. If the wind blew too fast, the machines had to be slowed down at just the moment when they would have been producing the most electricity.[6] But with its new power electronics, the 33M-VS was going to change all that.

The company's (in)famously confident management team was more than happy to tell investors that their new machine was going to revolutionize the wind industry. On October 12, 1992, the company held an initial public stock offering. Though they intended only to sell one million shares at about $10 a piece, they ended up selling six million shares and raising $92.4 million, which was supposed to head straight into continued R&D, production, and marketing of its new breakthrough turbine.[7]

Employees of Merrill Lynch, using the not-quite-kosher arrangement that became popular during the dot-com boom 'n' bust, acted as both investment bankers on the deal and analysts evaluating the company's merits. Unsurprisingly, they found their client's stock was a tremendous bargain. They forecast Kenetech's sales would grow 10,000 percent by 1995 and predicted that the stock's value would rise 50 percent in just a year.[8]

Other financial analysts were convinced, too, and the company's share price did skyrocket. At one point, Kenetech was valued as a billion-dollar business. With capital in hand and projects lined up, all they had to do was install the new wind turbines and start counting the money flowing to them from grateful utilities. "We're beyond Kitty Hawk and into the jet age," Edgar DeMeo, head of the Electric Power Research Institute's solar division, told the *New York Times*.[9]

THE LAST GREAT AMERICAN TAX SHELTER
OR THE INFANT GREEN-TECH INDUSTRY

On the day of Kenetech's IPO, a new issue of *Time* came out detailing the bizarre three-way presidential race that had developed between George H. W. Bush, Bill Clinton, and Ross Perot. Along with the political handicapping, the magazine ran a feature about the impact of environmental issues on the campaign as seen through the vice presidential contenders, Al Gore and Dan Quayle. In a down economy, so Quayle's story went, environmental issues had to take a backseat to job creation. Conversely, Gore argued that "sound environmental policies can be an engine of growth that will help the American economy compete with Germany and Japan in the 1990s."[10]

Although today our perceived competitors are China and the rest of the developing world, both political teams continue to hold the same basic positions. Breaking this stalemate gives technologies like the 33M-VS extra rhetorical value. If a clean technology is also a cheaper technology—or can be said to be a cheaper technology—it bypasses the political question. No one has to lose. "The Democrats argue that environmental decisions should be an integral part of economic planning," *Time* continued. "The Republicans seem to be saying the country should address environmental problems only when it can afford to."[11]

If, however, wind technology was as cheap as fossil fuels, both parties would be satisfied. But how do we get from expensive wind machines to cheap ones?

The world's leader in wind technology, Denmark, put in place a rigorous program to support turbine research, development, and deployment over the long term. They built research facilities that worked closely with the wind industry to both certify the quality of turbines and improve them. They created policies that enabled and prodded companies to share their learning so that the entire industry could improve, and they provided steady, long-term incentives for producing electricity from the wind.[12]

The United States took a more haphazard approach toward developing renewable energy. The most important piece of legislation was the Carter-era Public Utility Regulatory Policy Act, known almost exclusively by the acronym PURPA. The 1978 bill required utilities to purchase electricity from independent power producers that used renewables or generated both heat and electricity in the same plant.[13] But states were tasked with actually implementing the policy. Results varied. Most did little to nothing. However, California under Governor Jerry Brown was not most states.[14]

Having taken office in 1975 at just thirty-six years old, Brown's governance style was freewheeling and deeply influenced by the philosophical movements that infused the '70s. He openly supported E. F. Schumacher's call in *Small Is Beautiful* for an alternative economics and technological regime. Under Brown, California even got an Office of Appropriate Technology.[15]

Inside these broad strokes, Brown gave a lot of freedom to his fascinating band of underlings in the California bureaucracy to make things up as they went along. One particular PhD-packing philosopher of science, Tyrone Cashman, got to design the legislation that created what longtime California energy journalist Peter Asmus called, "America's last great tax shelter," otherwise known as the infant green tech industry.[16] "With Governor Brown's blessing . . . two diametrically opposed cultural camps—greedy exploitationists and well-meaning ecotopians—were brought together in a marriage of convenience that gave birth to today's modern wind farming industry," Asmus wrote in a 2001 book on the period.[17]

In those early days, tax incentives and guaranteed power purchase contracts were the wind industry's lifeblood. It was a classic case of what innovation researchers call a "demand-pull" policy, in which the lure of money is used to draw private enterprise to a field of technological interest.[18]

Incredibly generous tax incentives were doled out on the basis of how many kilowatts of wind capacity one could install. On top of the 25 percent federal tax credit one could already get, the California legislature gave renewable energy developers another 25 percent deduction on state income tax. Crucially, this was a capacity tax credit that was not based on actual electricity production but rather on how many towers a developer could get in the ground and have generate some (any!) electricity by December 31 of a given year. Returns were practically guaranteed.[19]

On the plus side, the policy drove the installation of a lot of turbines. After the incentives went into effect in 1981, California became the world's foremost laboratory for wind technology. Anyone with a tower and some blades flocked to lease land in three passes: the Altamont Pass east of San Francisco, San Giorgino Pass near Palm Springs, and Tehachapi, east of Bakersfield. In 1985 these three areas generated almost 90 percent of the world's wind electricity.[20] California had become the only real wind market on the globe at a cost of about $200 million a year in lost tax revenue.[21]

Cashman, who had spent time with the radical technology group the New Alchemists, didn't mind introducing a little chaos to jumpstart the transformation of the nation's energy system. He was—and remains—completely committed to natural systems.[22] But renewable energy was competing against fossil fuel technologies that had been supported for decades by the government and had fully matured. Something had to be done to create new wind machines. The high-profile Department of Energy research program had succeeded only in finding out what didn't work.[23] Perhaps, Cashman figured, government could employ other means to guide people onto a cleaner technological path. "I wanted to break the vicious cycle. There was no wind technology because there was no capital. And there was no capital because there was no wind technology," Cashman stated.[24] The tax credits certainly solved the capital problem. Money poured in from all over. From 1981 to 1986 at Altamont Pass alone, $1 billion was invested in 6,700 wind turbines.[25]

To their dismay, investors and wind companies discovered the machines were not ready for the industrial-scale task they had been handed. Suddenly, for the first time ever, there was a real race to improve wind electric technologies. Because of the tax credits, investors made money on the plants immediately, but the wind companies had to survive on the revenue from the sales. This means that the better and cheaper the machine, the more money they could make. Almost without a doubt, the California tax incentives accelerated the technical progress of wind machines all over the world by providing people who wanted to buy them. The machines got bigger and better, and their operators learned how to use them more efficiently. The cost of the electricity they produced fell by an order of magnitude. Without the tax incentives, imagining how that might have happened is difficult.[26]

The tax incentives had serious perverse consequences as well. As has been pointed out many times, if the government provides incentives for people to build machines, not produce electricity, they focus on putting the cheapest possible machine into the field as quickly as possible. As such, few of the early wind generators lasted much beyond the year in which they were installed. More subtly, because the wind farms had to be up and running by December 31 to get tax credits, the wind farmers' schedule was incredibly difficult. Come hell, highwater, or massive amounts of overtime, the projects just had to get finished. The haste required made a lot of waste, even for those well-meaning wind farmers who very much wanted to produce good turbines.[27]

The tax credits were "much too generous—ridiculously so" even by Cashman's own estimation, but he doesn't regret the legislative "stick of dynamite" that he threw. "What we did was make it so seductive that they would invest—even if the wind turbines didn't work."[28]

And at first, the wind turbines did not work.

Andy Trenka, who ran a turbine testing program for the Department of Energy at Rocky Flats in Colorado, claims that his team "couldn't keep the turbines running long enough to get good performance data."[29] And that was in the relatively boring and staid wind climate of Rocky Flats. In California, the wind was more capricious and more intense. Wind is a very, very local resource. Although the general direction and intensity of the wind can be culled from weather and climate models, the specifics of siting require understanding local topology and conditions extremely well. Even now, with all the science, data, and computing power we can muster, individual decisions about where a machine should go are still made on the basis of on-the-ground measurements conducted over the course of a year.[30] Understanding and modeling the wind's three-dimensonality is difficult. Imagine: It's like constructing a topology map that changes every second.

And, as noted before, because the power that can be extracted from the wind varies with the cube of its speed, small variations in the speed of the wind have an outsized impact on how much electricity can be produced.

The people who rushed into California's mountain passes did not have detailed wind knowledge from previous study nor did they have time to acquire it, especially given the conditions imposed by the tax incentives. And the bits of knowledge that companies like Kenetech (then known as U.S. Windpower) did generate from wind tunnel testing and experience was treated as a trade secret and kept from public consumption.[31]

An early Solar Energy Research Institute report then showed the surprise that analysts trying to work with the wind already felt. "Wind turbine developers/operators have been amazed at the enormous variability in energy production within the Pass, both between turbine arrays but especially within arrays," the report's authors wrote.[32] Within a single ranch, the quality of potential turbine locations varied tremendously. The best site had a wind energy potential two-and-a-half greater times than the worst. Yet they were separated by just half a mile and only sixty feet of elevation.

Wind energy assessments began to be carried out in the mid-1970s, but they returned large data discrepancies and methodological issues. The Department of Energy and the Pacific Northwest National Laboratory steadily improved the quality of data and maps. But it took a decade to provide high-resolution maps that were useful for more than the coarsest analysis.[33]

The net effect of all of the immature technology and lack of data was that—frauds aside—wind technology companies started out posting terrible early results. In the early days, the companies routinely overestimated how much wind power they could produce. Furthermore, machines broke down, pitched blades, and nothing worked as expected. A 1981 survey found the life span of the first generation of turbines was only *seven hours*.[34]

Kenetech was the first company to install a wind farm at Altamont on the last day of 1980. They pressed a machine into service christened the 56-50. It was a disaster. Of the 557 early models, an almost unbelievable 95 percent of them failed in the field or had to be shut down because of danger.[35] For that short period Cashman's plan appeared to be a dismal failure. Then, however, it started to work, aided by the California Public Utilities Commission's decision to come up with a series of standard contracts for PURPA-qualifying facilities. The Standard Offer 4 guaranteed wind farms payments pegged to fuel price projections that proved too high. The contracts became a boon to renewable energy developers like the solar company Luz and dozens of wind farmers. With a guaranteed price in play, the wind developers' profits became pegged to the spread between the cost of making electricity and an inflated sale price.[36] It worked like the feed-in tariffs that have driven much renewable energy development in Europe.[37]

There was a lot of money floating around for the company that could come up with a decent turbine. As a result, a lot of brilliant minds and smart technologists began to work on the problem of coming up with this better mousetrap. The competition got so fierce that the German turbine maker Enercon claims that the National Security Agency teamed up with Kenetech to steal their technology and patent it in the United States before Enercon could.[38]

With these forces at work, it's no surprise that the machines did get better. Kenetech's next turbine design, a hundred-kilowatt machine, was a major success. They were even as reliable as the Danish competitors, who had begun to make inroads in the California market. The secret of Kenetech's success, though, wasn't in their new machine's off-the-shelf characteristics; rather, the tight coupling of their operations center with their maintenance teams—aided and abetted by better sensors and data—kept their machines running. They learned by doing. The longtime wind observer Asmus wrote,

No other wind developer put so much of its resources into integrating the real-time computer controls of each individual turbine with immediate, on-the-ground technical assistance. The windsmiths who worked for Kenetech became a key part of the company's formula for success. They provided invaluable feedback about how each component fared under the very specific wind conditions and quirks of wind fuel supply at any one turbine site.[39]

Combining their new machine and operations model, they became the largest wind power developer in America and the Electric Power Research Institute's preferred partner in creating the fancy new variable-speed turbine.

But they were not the only wind company in America or the world.

Kenetech may have been the highest-flying and the most public with their ambitions, but both Danish manufacturers and the American company Zond were working hard to make slightly better turbines that would be very reliable. Vestas, a Danish company that now leads the world in wind turbine sales, had developed a rugged set of machines that weighed five times as much as their American counterparts.

Popular Science described the turbine, not quite approvingly, as "trusty as a tractor." The other turbines, more American designs, were light and high-tech. The message was clear: The Danes were agricultural and old world whereas the Americans were aerospacey and forward looking.[40]

The Danish-style machines might have been heavy and more expensive on paper, but they did not have the disconcerting tendency of flying apart in tough wind conditions. In the early years, they nearly always outperformed their American competitors—and they continue to lead the world. Cashman said the Danes "saved his butt."[41]

Furthermore, the Danish approach to engineering was fundamentally different from the American one. It was much more conservative, favoring incremental changes. Danish turbines had to pass rigorous testing to qualify for tax incentives in their country. The manufacturers did not want to mess around with reliability, so they made their machines sturdy. An academic report concluded that

> Gradually, practical and hands-on knowledge about the poorly understood technology accumulated. On the basis of this knowledge, the design rules were gradually improved. Design and development problems stemmed from turbine failures or from construction problems. The failures were often solved by making the turbines more solid, or, in other words, by "throwing metal on the problem."[42]

Innovation researchers also argue that the country's wind program privileged learning-by-doing over the kind of top-down scientific research program that the EPRI-Kenetech collaboration epitomized. The industry grew out of just a few local manufacturers before metastasizing into a larger industry supported by the government's testing services.[43]

The American companies getting smoked in the field were impressed by the Danish. The Risø test station established by the Danish government drew particular praise for its position in making turbines better. "The test station has played a major role in helping the Danish-built turbines establish an international reputation for reliability, and is widely regarded as a model of wise and productive use of government funds in promoting alternative energy," wrote one observer.[44]

Meanwhile, the American companies were betting on the Next Big Thing, the one development that would allow them to cut costs dramatically and unlock a huge, new market. "American designers constantly sought breakthroughs," Paul Gipe, a leading wind researcher, has written. "They wanted to bypass the drudgery of incremental development and bat a home run."[45]

Zond was an exception. Their heavily promoted 1995 machine, the Z-40, was basically a knock-off of the Vestas turbines that the company had been using for its projects. The company admitted as much with the popular press, noting that it was a Vestas "with a few improvements,"[46] that is, a simple three-bladed turbine that was heavy and built to face the wind. Because they worked, Zond built one on that design.

Instead of plowing money into a massive turbine R&D program, the company put its money into an extensive data gathering system codesigned with the Department of Energy. They created a special test stand set-up that recalled the testing rig that Thomas Perry used to design his famous "scientific windmill." A nine hundred–horsepower diesel engine subjected the wind turbine to forces equivalent to a hurricane. A shaft running from the engine to the gearbox pounded the components, inflicting thirty years worth of damage in just six months.[47]

Kenetech likewise sought fundamental changes in the cost structure of a wind turbine so they could hit that magical five-cents a kilowatt mark. Consequently, it had to be lighter . . . and capture more wind . . . and be more reliable . . . and less prone to other types of failure . . . and provide better power to the grid. That's a long list of tough problems.

To get the machines to look cheap enough on paper, Kenetech was forced to cut corners on reliability while simultaneously projecting a turbine life of thirty years.[48] Considering that few wind turbines had survived even a few years, let alone a decade, the estimate shows how desperate Kenetech was to say that it had a machine that would compete with fossil fuels without subsidies.

All across the wind industry, insiders were betting against Kenetech's new wind turbine. Jamie Chapman, a former vice president of the company, pointed out that "if there were an obvious answer" to wind's problems, "it would spread like wildfire." Presciently, Chapman warned that turbine efficiency was only one aspect of the overall performance of a machine. Operating and maintenance costs could swamp the benefits of getting the 10 percent boost from the variable speed rotor.[49] Maybe if they changed a turbine too much, it would break.

BUBBLE BURST

On July 29, 1994, Hank Hermann, a wind energy consultant and analyst, visited the new Buffalo Ridge wind farm in Minnesota, which was stocked with brand spanking new 33M-VS turbines. The company was still riding high, with that $1 billion valuation and what the company described as $1.7 billion in backlog contracts with utilities. Not only that, but any day, the magic turbine was going to level the playing field with fossil fuels once and for all.

Walking the barren land on a beautiful warm day, Hermann inspected the machines to see if they were, in fact, the promised Five-Cent Dream Machine. What he saw caused him to send out a note to his investment clients that basically screamed, "Sell! Sell!" The ground was littered with blades and cracks near where the blades attached to the hub were clearly visible. Sixteen of the seventy-two turbines (22 percent) were not operating, even though the wind was blowing at speeds that reached fifteen miles per hour. It was not a pretty scene. In an August 22 note, he wrote,

> Admittedly, it is possible the Buffalo Ridge availability factor at the time of the visit was an aberration. However, while standing on the road by some of the operating wind turbines, the loud groaning, clanging, whining noises emitting from several of the machines strongly suggested to my untrained unscientific ear that meaningful problems may exist with some of these machines.[50]

Kenetech hit back at Hermann immediately, citing his affiliation as a consultant with New World Power Company, a competitor. Their responses only bought them a little more time.[51]

Then, *Windpower Monthly*, a leading trade magazine, put out an extensive feature detailing problems with the new turbine in its

September issue. It put the 33M-VS troubles Hermann saw in perspective. One of Kenetech's suppliers confirmed trouble with at least one component, the hydraulic boxes. Eyewitness reports from a Palm Springs wind farm found the situation there comparable to what Hermann had reported from Buffalo Ridge. "Talk throughout the wind industry was of most of the 33M-VS blades in Palm Springs being cracked or damaged, of times when most 33M-VS turbines in Palm Springs were apparently shut down when the wind reaches about 35 mph," the trade magazine wrote.

Kenetech maintained that any problems were just normal launch issues, but that didn't jibe with what the company had been saying for years about the fabulous new machine. The 33M-VS was supposed to be *less* prone to failures thanks to its variable-speed rotor, and yet nearly every piece of the turbine was falling apart. "They've done just about every major component and I don't see that as a teething problem," said an anonymous wind expert quoted by *Windpower Monthly*. "You might have technical problems, glitches or control problems, but you don't have blades coming apart at the seams or losing whole generators."[52]

A different anonymous source noted with a hint of snark, "If you spend $69 million on developing this cosmic technology, you shouldn't have these problems."[53]

The 33M-VS, eventually renamed the KVS-33, obviously never competed with fossil fuels. Years later, wind industry insiders were still referring to the "Kenetech fiasco."[54] Wind electricity prices, on average, did not drop below the magic five-cent threshold until 2002.

But maybe hitting that threshold wasn't nearly as important as the company made it seem in order to differentiate its product. Electricity markets and wind farm construction were considerably more complicated than Kenetech made it seem. Regardless, wind power companies started to become profitable several years before that.[55] Turbines weren't everything, even if they were the sexiest part of the industry.

The mechanical problems probably sealed Kenetech's fate, but the company also encountered stiff resistance from environmentalists who argued that they knowingly installed turbines that would kill federally protected birds in Altamont Pass. In retrospect, in regard to the local habitat, the pass was one of the worst places to site a wind farm. Although Altamont turbines killed thousands of birds in its early years, other wind farms, even the other early California areas, have shown much, much lower death rates.[56]

Further, persistent concerns about the aesthetic impact of wind turbines exacerbated the public relations problems at Altamont. People thought the wind farms were an eyesore. The early wind farms, though, might not be a good indicator for the overall social acceptability of wind machines. In the early wind areas, all kinds of different turbines were put into place; there was no uniformity of design. Broken ones were left standing, junking up the place. Spare parts were left lying around. So the wind farms were not good neighbors, perhaps because many were not expected to be in business for very long.[57] Concerns that the landscape had lost its scenic nature were equaled by problems with the human elements of the seemingly inefficient wind farms. Landscape researcher Richard Thayer wrote of the Altamont area,

> When wind is blowing and most other turbines are spinning, large numbers of non-operating turbines apparently lead to public perceptions of technological or managerial incompetence or tax fraud. Inoperative turbines seemed to equal or exceed siting, design and scenic character in causing negative attitude formation among subjects.[58]

Overall, the bird kills, landscape transformations, and incompetence of the early wind farmers made getting later farms sited a nightmare for Kenetech, even though they were not the worst offenders on any of these items. They were, however, the face of the industry, so they bore the brunt of public concerns. Prospective real estate developers with their eyes on building exurban rural estates were loath to see their subdividable properties overshadowed by windmills, even though subsequent evidence suggests that wind turbines do not depress housing prices.[59] As a result, projects proceeded more slowly than anticipated.

Hoping to get their turbines working and their projects in the ground, Kenetech limped through 1995 with their share price in the tank. The end came when Kenetech filed for Chapter 11 bankruptcy on May 29, 1996. A company that had ridden California's tax incentives to the pinnacle of the industry, weathered their repeal, and developed the most expensive turbine in history, had been felled. Their old rival Zond bought their patents and research on the variable-speed rotor on the cheap and developed a very successful turbine based on the technology.[60]

The company's failure speaks broadly to the issues of renewable energy technology. Most obvious is that breakthrough technologies are very difficult to develop. Although virtually all wind machines now use variable-speed rotors because they *are* a good idea, Kenetech tried to make too many changes too quickly in hopes of dramatically reducing the price of wind power. What they ended up with was a paper turbine that broke regularly under real conditions.

Danish wind turbines, by contrast, rarely tried to make major changes to their designs. They scaled up slowly and incorporated few new features.[61] This was partially because the government mandated that they submit to reliability testing before they could receive domestic subsidies. However, the business culture of the United States contributed as well, said Fort Felker, who worked at Kenetech from 1994 to 1996 and who is now the director of the Wind Technology Center at the National Renewable Energy Laboratory. Felker said,

> We felt that we had a huge technology head start and we had no interest in collaborating with anybody. I don't know that the Danish model could have worked for the United States. The Danish environment had strong government support of the industry whereas in the US we didn't. We were struggling to stay in business and make wind be successful in light of dirt cheap fossil fuels.[62]

Danish lawmakers actively pushed the businessmen who bankrolled their technologies to be familiar with how those technologies worked, and the government strove to create innovation networks that bound researchers, people working in the field, and corporate types as closely as possible.[63]

However, the gusty policy environment of the time, with incentives swaying to and fro, made it difficult for companies to count on anything but technology. American companies were constantly pushing the pace of technological development beyond what made sense because they never knew when the playing field would shift. In fact, it was only the long-term Standard Offer 4 contracts that allowed the industry to grow in the first place. But perhaps the wind companies' experiences in the passes of California should have taught them a different lesson about how to reduce the cost of energy they produce: Operations matters. In 1992 the R&D manager of PG&E, which bought most of the wind power in the world at that time, noted that half the cost reductions on the wind farms in California "have arisen from on-the-ground experience in building the machines, installing them and keeping spares."[64] What's more, as noted in *Time* in 1992,

Wind's success says something about a dicey political issue: Should government tamper with free enterprise to nurture a new technology? The answer for renewable energy sources is definitely yes. Had manufacturers and utilities not received state and federal assistance early on, the future of wind power would now be controlled by either Japan or Europe; both have consistently funded wind research.[65]

The problem is that American policymakers couldn't make up their minds. Businesses lived in constant fear that their tax incentives or other support would be cut by capricious lawmakers, which made it harder to make sound long-term decisions. Congress has confirmed the industry's worst fears: The production tax credit, the main governmental support for the wind industry, has been extended for arbitrarily short periods of time for almost two decades. It was allowed to lapse in 1999, 2001, and 2003. Those gap years devastated domestic wind farmers.[66] It's not too much of a stretch to say that this is the worst way to dole out subsidies one can imagine and the one strategy least likely to lead to a healthy industry. The oil industry, by contrast, received a huge, steady tax break from 1913 until 1975, despite decades of criticism. Neither the break nor its abolition, in the words of Harvard's Richard Vietor, was tied "to any rational concept of energy policy,"[67] but the oil industry was powerful and its lobbyists got what they wanted until the energy crises of the '70s. The wind industry has never managed such consistent success.

So whereas the American turbine manufacturers twisted in the legislative wind, Danish policymakers had no problems supporting their homegrown businesses and promoting their potential as an export industry. Even during downturns for the Danish wind industry over the last thirty years, the government has proactively worked to keep knowledge within the industry. "The Danish have preserved knowledge that otherwise would have disappeared or become worthless. They codified it in reports and kept a lot of the same people employed in the industry for 20 or 30 years. That just doesn't happen at all in the United States," said Gregory Nemet, an energy innovation researcher at the University of Wisconsin. "There is a real benefit to stability and avoiding this depreciation of knowledge."

Like human archives, experienced hands help keep mistakes from being repeated. They have contributed mightily to the success of the world's leader in wind energy, the Danish company Vestas, which employs twenty thousand people and generates six billion euros of revenue annually, largely from exports. So the lessons from thirty years of wind power development aren't sexy or particularly shocking, but they are the kind of common sense that American energy policy needs.

First, technological breakthroughs, politically appealing as they seem, are rarely as easy to achieve or important as advertised. Second, cost reductions are as likely to come from learning by doing as by designing paper tiger machines. Third, consistent incentives are more efficient than the yo-yo versions that have typified American policy. Fourth, as shown by the Danish experience, creating high-functioning innovation networks requires more than tax incentives. Government support for R&D and reliability testing are essential. Fifth, during the ups and downs of the business cycle, knowledgeable workers need to be kept within the field.

For too long, we have suffered from an Edison complex, delusionally believing that some superhero inventor will solve the energy problem. The truth is that energy technologies don't work like that. The development paths are grinding and long because the levels of reliability required are high. If governments are going to invest in them, they have to be willing to act consistently and for many years at a time.

In February 2009 the production tax credit for wind was extended for three years, and given the economic tough times of the period, this allowed credits to be converted into grants. Without the moves, which were included in Barack Obama's stimulus package, the industry might have collapsed again. "The stimulus package essentially saved the renewable-energy industry in the United States," a venture capitalist told *The Atlantic*.[68]

Perhaps American energy policy is finally growing up.

chapter 25

Energy Storage and the Return of Compressed Air

N 1991 A SMALL POWER PLANT opened up in McIntosh, Alabama, operated by the Alabama Electric Cooperative, one of the old public utility companies that brought power to the hinterlands of the deep South.

There was little fanfare. No one reported from the gates of the plant, which is located in a rural area about forty miles north of Mobile. It wouldn't have been much to look at on the surface anyway. The plant's footprint is about the same as the small high school that sits across Jefferson Davis Highway, if we include its football field and baseball diamond. A chemical factory in the area dwarfs the plant.

What matters about the McIntosh project is not so much what's happening where you can see it but actually the action deep underground. It sits atop a large salt dome, the kind of geological structure that was an early target for wildcatters looking for oil. Petroleum and natural gas tend to get trapped underneath these impermeable rocks: Poke through the layers of salt and sometimes people can slurp up some hydrocarbons. A gas company explored this particular dome in 1945, and when it came up dry, Olin, the chemical manufacturer, purchased it, using the salt for various purposes.[1]

For decades, the dome lived a relatively quiet life until 1987, when two small test wells pushed down into it. Geologists looked at the rocks that came back up and decided that one area had salt that was perfect for what they were planning: solution mining. They then pumped water down wells, which dissolves the salt, then pumped the briny water back up. What's left over is a cavern underground. For every fifty gallons of water they pumped down, they made one cubic foot of cavern. And the best part is that the salt heals itself up, becoming impermeable again.

In October 1988 an engineering contractor drilled a hole that reached 2,400 feet underground. Then the water pumping began. The briny water got passed off to Olin for use in making chloralkali. The cavern grew. By November 1990, the cavern ran 900 feet long and had a maximum diameter of 238 feet. It had a cubic volume of 19 million square feet. It is shaped roughly like an ear of corn inserted into the ground tip down.

What would anyone possibly want a corn-shaped cavern underground for? To store energy. The cavern acts like an underground pressure vessel, storing up power for later use. During off-peak times, when electricity is cheap, they can run that electricity to air compressors, which pump pressurized air into the cavern. Then, when electricity is expensive, they flip the switch and run that compressed air through a turbine with some natural gas. They have effectively stored the electricity to be used for later. The current state-of-the-art technology uses the wind electricity to reduce the amount of fossil fuel that's burned in the generator by 60 percent, and future designs hope to eliminate combustion altogether.[2]

McIntosh and its salt dome hardly seem like a key destination on the map to the future of renewable energy in America. But they may be. In 1991 the plant above the cavern went into operation—and has been humming along ever since. Executed purely with private money for economic reasons, the success of the plant provides hope that intermittent power sources like wind and solar will one day soon be able to provide power twenty-four hours a day, seven days a week. Compressed air energy storage plants may be the key technology that allows wind to compete with baseload coal power.

Wind power, as David Marcus, president of General Compression, a new energy storage company, likes to say, is intermittent on every time scale: "It's a fractal problem." But, he says, people have approached the technological conversion of wind into electricity all wrong. "The problem is that the current technology has turned the intermittent resource into intermittent power," Marcus said. "That's a technology problem, not a resource problem."[3]

In fact, the water pumping windmills of the nineteenth century solved the intermittency problem in their time. The mills were used to pump water into storage tanks, where it could be used to grow crops or water cattle. Wind energy was effectively converted into water, which was then converted into food energy. It was a long chain of transformations and surely not the most efficient system, but it worked, and the millions of mills that dotted the country were key to the settling of the interior of North America (see Chapter 6).

Marcus's company, located outside Boston, near where the first self-regulated water-pumping windmill was invented, has a new technological solution. It's still under wraps, but General Compression received $16 million in venture capital in March 2010 to bring their new compressed air storage system to full scale. It will use no natural gas and has been designed exclusively for use with wind plants.[4]

But the current leader in the field can be traced right back to McIntosh. Energy Storage and Power Corporation has two big grants from the Department of Energy to build next-generation compressed air plants with utilities in California and New York. Michael Nakhamkin, who designed the McIntosh plant, is ESPC's founder.

In almost any scenario in which renewable power dominates the future, cheap storage will be a necessity. Dozens of companies are now offering flywheels, different types of batteries, and other even more experimental enterprises. Many of them are new to the business and many questions remain about them.[5] Whereas battery technology is filled with nanotechnological whizbangery, compressed air storage is nearly the opposite. It's a well-known entity, not just because of the McIntosh plant but also because it's actually been used industrially since the late nineteenth century. One 1890s mechanical engineer, Robert Zahner, was enthusiastic about its benefits:

Compressed air is the only general mode of transmitting power; the only one that is always and in every case possible, no matter how great the distance nor how the power is to be distributed and applied. No doubt as a means of utilizing

distant, yet hithero unavailable sources of power, the importance of this medium can hardly be over-estimated.[6]

For a brief moment, it seemed as if the future of energy transmission
lay with compressed air. What a different energy system that would have
been for renewable energy! Compressed air, unlike electricity, can easily
be stored. The ragged intermittency of renewables could have been
cushioned by caverns of air. This remarkable capacity was not lost on
Zahner. "Compressed air is also a *storer* of power, for we can accumulate
any desired pressure in a reservoir situated at any distance from the
source, and draw upon this store of energy at any time," he wrote.

Electricity has some fantastic properties, but it's also the strangest of
commodities. For one, it is consumed nearly instantaneously. We call
wires "live" for a reason. As it turns out, storing electricity has turned
out to be one of the most difficult challenges of the last one hundred
years. The problem has confounded every chemist and electrical engineer that has taken it on, from Edison to today's lithium-ion jockeys.
There has been no saving electricity for a rainy day.

So engineers came up with logical workarounds to sell their product. They built the entire energy system around balancing load and demand. Most plants run most efficiently when they're left on all the time,
so it became important to smooth the amount of light and power used
throughout the twenty-four-hour day, even if people didn't need it. As a
result, the engineering properties of the turbines and electricity drove
the creation of entire social systems. When the traction companies had
excess nighttime capacity because they ran more trolleys during the
day, they built amusement parks at the end of the line and extravagantly
lighted them to suck up the extra juice. Utilities also invented (wasteful)
uses of energy like electrical resistance heating to help balance the increase in load from air conditioning. Thus, the inability to store their
product has shaped the entire industry from top to bottom.

In modern systems, entire classes of power plants are switched on
only a few days a year, when they make some of the dirtiest and most expensive power. They're like fireworks stands that remain empty all year
and then make all of their cash in the two weeks around the Fourth of
July. The electricity business is the ultimate just-in-time delivery game.

Conversely, compressed air has completely different properties. Air is a mix of gases. It might not be quite as easy to store as a liquid like oil, but decades of natural gas storage have proved that it's possible to do so—and cheaply. At any given moment, something like 1.7 trillion cubic feet of natural gas are stored underground in the caverns and sandstones of America.[7] That's enough to fill 1.1 billion cargo containers, or the Sydney Harbor sixty times over.

In the '90s, after the McIntosh plant proved that compressed air storage could be done, there were high hopes that many utilities might jump on board and build their own. "We expect the CAES plant technology pioneered in Alabama to lead to widespread application in this country," said Robert Schainker, the manager of the Electric Power Research Institute's Energy Storage Program. "Three fourths of the United States has geology suitable for underground air storage. At present, more than a dozen utilities are evaluating sites for CAES application."[8]

But they didn't. Nakhamkin said that the tiny Alabama Electric Cooperative was in a unique position. During the day, they had to purchase expensive peak power and run their coal generators, but at night, they had to switch off their power plant, which is very inefficient.[9] They needed something to suck up nighttime power. Unlike bigger utilities, they couldn't balance their loads as well. Other utilities could have saved money by building similar plants, but it was not "critical savings," Nakhamkin maintained. "Rich people don't talk about how to save five dollars or ten dollars."[10]

Because the McIntosh plant was specifically optimized for the situation in Alabama, it was not very flexible. It could take power from the grid only in 50 megawatt chunks, for example. Its startup time was also long, which decreased utilities' interest in the technology.[11] Nakhamkin said that his next generation designs will provide plants that can fit more utilities' needs.

But what's really brought CAES roaring back is Zahner's "distant, yet hithero unavailable sources of power." In the United States, that's turned out to be wind electricity, which has led alternative power additions to the grid for the last several years. Quick-to-deploy, cheap, and with less carbon risk than other alternatives, more wind than coal is slated to be added to the grid from now until 2030.[12]

The importance of compressed air storage is not only that it would make it easier on utilities to accept gigawatts of wind power on their transmission grids, but also that it would allow wind to compete directly with coal. To balance the amount of generation and load on the grid, operators decide to switch on power plants in an order determined by the price at which they can produce electricity. Although a kilowatt of wind is currently cheaper than the average wholesale electricity price, it only tends to displace natural gas peaking plants. Wind doesn't displace baseload power because the nuclear and coal plants can't ramp up or down quickly.

The latter's power is sold in blocks with the expectation that the plant will be running 90 percent of the time. No current renewable technology can provide that kind of power. The only way that wind is going to displace these coal power plants is to offer the ability to provide power nearly all the time.[13] To do that, they'll need storage, and most current estimates from inside and outside the electric utility industry point to compressed air as the cheapest option out there. "CAES is the least cost utility scale bulk storage system available. If other factors such as its low environmental impact and high reliability are considered, CAES has an overwhelming advantage," one Department of Homeland Security analyst concluded.[14]

Four major projects are now under way across the country. First Energy, a large utility in the Ohio area, recently purchased a long-awaited project in Norton, Ohio. This project could store over two gigawatts of power in an abandoned limestone mine. An Iowa project is also moving off the drawing board. That project is particularly important because it will explore pumping compressed air in porous sandstone just the way natural gas is stored. In California, PG&E received $29 million in matching funds to build a three hundred–megawatt plant in Kern County. New York State Electric and Gas received the same amount for a similar facility in the town of Reading, New York. It will use an existing salt cavern there.[15]

Although Nakhamkin and General Compression have made incremental breakthroughs in how their compressors work and integrate with existing facilities, compressed air technology is not likely to experience massive changes in its cost or attributes. The plants will get a little cheaper, more standardized, and better, but scientists are not likely to stumble on a huge cost surprise, good or bad.

Maybe that's its strength. No technology is an island: Like citizens of a nation, technologies are linked and woven together in sometimes surprising ways. The compressed air storage plant that enabled a tiny Alabama utility to store its cheapest, dirtiest coal power may now enable a breakthrough energy *system* that provides cheap, reliable wind power to the world.

Compressed air storage may end up providing a powerful example of what sociologists Raghu Garud and Peter Karnøe call the "bricolage" path of technological development.[16] While materials scientists pursue battery breakthroughs, compressed air energy storage systems will mash up existing commercial parts into an innovative solution to the new problem of intermittency.

chapter 26

"Throw Software at the Problem"

I N 1841, SOMEWHERE NEAR LOUISVILLE, techno-utopian John Etzler was trying to burn bushes. Wandering what was then still the West, he was tinkering with the prototype for what he hoped would be society's new democratic power source.[1] The would-be prophet had a vision for reinventing society—harnessing the power of the wind, waves, and rays of sun beating down on Kentucky and everywhere else.[2]

Perhaps radical change was in the air. The Industrial Revolution was slowly infecting the rest of Europe, and up in New England, capitalists were tapping the power of falling water to power textile mills run by gangs of young women. Etzler himself had just completed his stint in New Harmony, Indiana, where Robert Owen purchased a town from the radical Christian utopianist, George Rapp, to create a socialist utopian settlement.

We can imagine Etzler, short and stocky with a head that one of his contemporaries called, "massy, the brow strikingly protuberant" from the "large volume of *brain*" inside it, sweating in the sun. Intense, with plenty of "nervous energy" in his eyes, he must have spent hours arranging a set of mirrors with the hope that he could concentrate enough sunlight to catch the brush on fire. It would be his proof-of-concept that he could drive a steam engine with nothing more than reflected sunlight.[3]

The principle is well known to any school kid who has used a magnifying glass to burn ants. It goes like this: The sun's direct rays can be collected over a relatively large surface and focused onto a point. Make your lens or mirror bigger and the point gets increasingly hotter.

Variations on the burning mirror theme had captivated a certain kind of aristocratic mind for millenia. Passed down through a clan of dedicated polymaths from Archimedes, who purportedly used them to set fire to the Roman fleet besieging Syracuse, to DaVinci, to the seventeenth-century Jesuit brainiac Athanasius Kircher, the secret of how to focus the weak power of the sun into an intense heat capable of burning a wooden navy was just emerging from legend and myth into the more prosaic realms of a newly science-obsessed Europe.[4]

But how could one make a burning mirror of sufficient size to drive a steam engine or smite your enemies? Large amounts of shiny material in a concave shape were not easy to come by—and it was heavy. In the early 1800s a mirror smaller than the one over a bathroom sink weighed 110 pounds.[5] Some enterprising nobles broke their materials into segments and attached them to concave wooden scaffolding. That yielded mirrors with diameters up to five feet. Etzler, however, had a different idea. He wanted to use whole arrays of flat mirrors focused on a large boiler. In his 1833 manifesto, *The Paradise Within Reach of All Men, Without Labour, By Powers of Nature and Machinery*, Etzler wrote,

> It is immaterial, too, of what size, form, or colour the pieces of such a mirror be; they are all to be of a flat surface. There is no curvature of their surface required as in the usual burning mirrors. All that is required for producing a focus, or burning spot, where all the reflections are concentrated, is to give each flat piece of such mirrors its proper place and inclination towards the sun.[6]

The question of how, exactly, to keep those mirrors in their proper place was easy. It "requires no laborious computation or preparation," he wrote. They could be kept aligned with "nothing more" than "moving the mirror to the sun's motion for casting its concentrated reflection or focus always on the same spot." Etzler was a great futurist, but he was never much of an engineer.

Tracking the sun is perhaps the toughest problem solar engineers face, which Etzler might have known if he had read about Léon Foucault's Heliostat or one of the other devices that eighteenth-century thinkers built to track the sun in natural sciences experiments. It was not until the early 1840s that any type of device to point the sun's light in a particular direction was available. And Johann Silbermann's heliostat could move only a tiny mirror, suitable only for microscopic investigations. By 1862 Foucault's was bigger and better, but it, too, was nowhere near the scale required to point a bunch of mirrors at a boiler to make it produce steam.[7] When a Greek engineer tried to replicate Archimedes's legendary mast-burning feat in 1973, it took seventy-three perfectly coordinated men holding flat mirrors to set a wooden ship aflame.[8]

How far Etzler got in his experiments is unknown, but we can almost be sure that he failed, leaving only the idea behind. It waited to be rediscovered for more than a century.

The good news is that sun tracking is a data and control problem in an era defined by new advances in data collection and precise control of large systems.[9]

And it was there that a successful Internet entrepreneur-turned-investor named Bill Gross began to reimagine solar power plants for the twenty-first century. How to keep the mirrors pointed in the right spot with the least effort led his company, eSolar, down a remarkable, data-driven path that led Google to make the Los Angeles startup one if its first two energy investments.[10] Gross unknowingly echoed Etzler's thinking in deciding to make small, flat mirrors that would huddle close to the ground, reducing the wind's effects. In fact, the mirrors are so small—less than two square meters—that they are "self-ballasting," meaning that they aren't actually screwed into the ground. Their diminutive size also means they can be manufactured in a factory and shipped to a project.

All these ideas may have made sense to previous engineers, but when the mirrors get that tiny, the enormous numbers of mirrors need to be controlled precisely. A single forty-six-megawatt eSolar modular unit uses a field of 200,000 mirrors. By fossil fuel standards, even a modest-sized plant of 230 megawatts would use one million mirrors. In the past, that would have been a daunting software control problem.

Every single mirror needs to be independently pointed to keep the plant working. Gross said that

> in the last decade, there's been a 1,000-fold increase in computational power, so now we can put a $2 microprocessor in every mirror and it costs almost nothing—about one and half percent of the material cost. So every mirror that is tracking the sun during the day has its own computer. And the computational power of a microprocessor today is mind-boggling. It's a 16-bit microprocessor with eight I/O ports. It's like an IBM AT (PC) in every mirror—that was a $5,000 computer in 1985. This completely wouldn't be possible without Moore's Law.[11]

In the classic formulation, Moore's Law holds that the number of transistors on an integrated chip would double every eighteen months. Though the definition has been extended and squished and morphed since Intel chief Gordon Moore first brought the idea into the mainstream in 1965, the basic storyline still holds.[12] As Gross likes to say, commodity prices are headed up in the coming years whereas "CPU power per dollar" continues to get cheaper. He's structured eSolar to take maximum advantage of cheap computation. They should have plants running in the United States, India, and China in the next few years, and all of them will be running the same algorithms. Improvements in software in any of the plant locations will be portable to its cousins overnight. "Now that we see and saw that software is the solution, we push for it even harder to see if you can make a power plant that has a larger software component and less steel," Gross concluded.

By applying information technology to energy problems, eSolar is one company among many new venture backed–startups that has realized that they can achieve major breakthroughs. To twist the old Danish wind mantra that many problems could be solved by "throwing steel on the problem," these entrepreneurs are learning how to throw software on the problem. Beyond the code first, Gross and his ilk are also bringing a new sensibility to energy space. Silicon Valley technologists live for disruption. Weaned in the computer industry, their engineers do not have the same sense of technological aesthetics or traditions that engineers who work for utilities and power companies do.

The software and computer entrepreneurs who are pouring into green tech bring into the energy space fundamentally different ideas about the way technology and organizations should work. Historian Thomas Hughes identified the ways in which what he calls the second industrial revolution—closely linked to electrification—differs from the information age. "Hierarchy, specialization, standardization, centralization, expertise, and bureaucracy became the hallmarks of management during the second industrial revolution," Hughes wrote. "Flatness, interdisciplinarity, heterogeneity, distributed control, meritocracy and nimble flexibility characterize information-age management."[13]

Silicon Valley, in particular, has a fundamentally different culture from the industrial ones that preceded it. The Valley is into "information sharing, collective learning, informal communication," and the companies it produces are usually fast-moving startups who are flexible and networked.[14] It is sometimes hard to pin down how business culture influences the technologies an industry produces, but we know that it does. Certain solutions *just make sense*.

Take the size of power plants, for example. Conventional wisdom has been to build the biggest plant possible in order to wring a little more electricity out of the plant's steam. Bigger plants technically work better, so fossil fuel plants grew through most of the twentieth century. The biggest plant in 1903 could put out five megawatts of electrical power. By the 1970s the largest plant stood at 1,300 megawatts—growth of 26,000 percent.[15] Countercultural environmentalists reacted against the increasing centralization and scale, arguing for the deployment of millions of small devices to make power closest to where it's used (see chapters 18 and 19).

The company eSolar settled on a smallish but modular plant size: forty-six megawatts. The odd size shows that Gross wasn't picking ideas off the shelf; instead, he rethought the requirements of power plants in today's market and compared them with the types of business factors that would impact the company's ability to deploy as quickly as possible. They optimized for speed to market as much as for efficiency.[16]

The Federal Aviation Administration requires that towers over two hundred feet go through an arduous permitting process, so that became the hard limit for the tower's height. In addition, eSolar wanted to have a smaller footprint so that they could locate their plants closer to transmission lines in order to avoid the problems that companies like BrightSource are having with environmental opposition out in the Mojave desert.

Some of eSolar's competitors thought they were nuts to build such small plants. Bob Fishman, CEO of a solar thermal competitor, Ausra, dismissed eSolar's approach. "I've looked at it, and I can tell you right now, there's a direct correlation between size and cost. If they want to build 30-megawatt plants, they can have at it," lectured Fishman, a longtime utility executive in 2008. "They're not going to change the laws of physics."[17]

But it turns out, building the plants around their business model instead of the other way around might have been a smart decision. Other companies have struggled to bring their demonstration plants online, even if they've been able to land big contracts with California utilities. Because eSolar needs only 250 acres per plant, it's easier to snap up and build on land that's been used for some other purpose like farming. As a result, eSolar is now a clear leader in the solar thermal space. In 2010 eSolar landed a $5 billion, first-of-its-kind deal to build two gigawatts worth of power plants in China and a one-gigawatt deal in India.

Ausra, for its part, laid off employees in early 2009 and was purchased by the French energy giant Areva in 2010.[18]

THE DREAMSCAPE OF INVENTORS

Computing power won't just change the operation of renewable energy plants. Software has transformed fundamental scientific research in the underrated branch of chemistry known as materials science. The history of energy technologies is built atop the history of the materials that were available for their construction. The characteristics and performance of materials like iron or glass, as determined by their molecular architecture, are the raw materials on which inventors and engineers imaginations' feed and build. As no small number of historically oriented materials scientists have pointed out: Materials have been so important to civilization that we've tended to class epochs of humanity (the Bronze Age, etc.) by their use.[19]

The introduction of steel, goosed by materials scientists like Henry Bessamer, underpinned the creation of today's turbine and internal combustion engine–heavy energy system. The development of better steel alloys, which allowed the construction of bigger and hotter running power plants, was a major component of the tremendous drops in the cost of producing electricity over the last century.[20]

Whereas the increasing efficiency enabled by turbines wiped out most alternative energy technologies, the development of plate glass allowed the sun's heat to be harnessed on the cheap. As discussed in chapter 12, the creation of double-paned windows in the 1930s opened up new design possibilities for solar houses. Coated glass types that let in light but trap heat made even more energy-efficient housing possible. Materials science was also what created the silicon-based technologies that drive computers and spun off the first tolerably efficient photovoltaic cell.[21] Grinding materials engineering has led to the tremendous advances in PV systems since then.[22]

Beyond photovoltaics, batteries and energy-scavenging devices to convert heat into electricity would both benefit from better materials. Wind turbines could get stronger and lighter. Building materials like cement could become less carbon intensive.[23] To say materials science holds the potential for transformative change across the whole greentech spectrum is no exaggeration.

But there's a problem.

Materials science remains substantially a trial-and-error bench science. If we want a material with certain properties—say, a much better battery—then we have to just keep trying different synthesis techniques to come up with new options. We test them to see what worked and then keep iterating on that process. Getting a material from the drawing board out into the field takes decades. Even once we have a material, creating the manufacturing process to make a lot of it involves years of trial and error.[24] In 2008 Gerbrand Ceder, an MIT materials science professor, and other researchers wrote,

> Searching for novel materials is typically a rather random and therefore somewhat unpredictable process. The initial focus on new materials is usually based on a single outstanding property. Hence, attempts towards scale-up and commercialization start

without access to other key properties which affect whether a material can be commercialized: reliability, stability and degradation, processability, etc. Deficiencies of a material in any of these areas can throw a materials development program off track.[25]

In other words, materials scientists find a material that's good at converting sunlight into electricity and start trying to make more of it. When they do, they discover all sorts of little quirks. As the years go by the once new, good-looking material starts to seem more like a drunk, unstable husband. If only scientists were able to predict which materials would be a good match over the long haul.

Ceder thinks he can do just that. The idea came to him in the waning days of the dot-com boom. He was doing some sidework with a colleague for a startup that used algorithms to predict consumer movie preferences—not unlike Amazon's famous recommendation engine. After a few months he returned to his normal work of thinking about compounds and their various properties like strength, conductivity, and the ability to store electrons. When he did, suddenly all the meaningless work for the Internet startup took on new significance. "I remember sitting down and going, 'Oh my god, this is just like movies: The people are the compound and the movies are the properties,'" Ceder recalled. "I really got that wacky idea there at the startup. It's the only good thing that I got out of it. I think they sent me a hat, too."

The idea, developed with Dane Morgan, now a materials scientist at the University of Wisconsin Madison, grew into what Ceder calls the Materials Genome Project. It is no less than an attempt to solve the quantum equations for nearly every inorganic substance that humans can produce. There are something like fifty thousand inorganics out there, Ceder estimates, and maybe another fifty-thousand left to be discovered. Over the last couple of years and just in Ceder's lab with half a million dollars of computers, they have quantified the properties of thirty-one thousand materials. "We already have enormous coverage of the known universe of materials," Ceder says. "We're showing that this whole discovery process is about to be accelerated 1000-fold."[26]

Ceder's team isn't necessarily blazing a new scientific path. Scientists have long been able to predict some properties about materials from the principles of quantum theory. By solving the equations of quantum mechanics, they can know how a material is likely to behave. "What's new is that we have automated that," Ceder said, "and automation allows you to scale." Going back to his Internet experience, he said that it wasn't the Web, per se, that brought us the wonder of the Web. Rather, it was the automation of information collection by Web crawlers that has made the universe of data accessible to human designs.[27]

The difference in materials science is that someone has to make the stuff once they've found it. Synthesizing a new material is difficult. No matter what, there is a lot of trial and error involved as we try different chemical treatments to find the right pathway to transform a set of parts into a new material. The worst part is that if we make something and it doesn't work the way we thought it might, what happened is not always clear. "You don't know if it doesn't work because it's not supposed to work or because you did something wrong," Ceder said. But computation will help there, too. Ceder's team knows what the material is supposed to do in an ideal world, so if it's not behaving right, they can look at their synthesis for the problem.[28]

The Materials Genome Project focused first on new energy storage materials. Some of Ceder's teams' first results were a better lithium-ion battery that made its way to the pages of *Nature*.[29] They launched a photovoltaics program in early 2010 and may devote some time to thermoelectric devices, too. "People should realize that we are on the cusp of this scaling revolution," Ceder concluded. "Discovery is going to be done in a very different way."

And it's all thanks to the combination of huge amounts of computational power paired with the data mining knowledge to take advantage of it.

SENSING THE ENVIRONMENT

Consider the dam. A dam transforms a river into a water battery. An ephemeral energy flow becomes a liquid building ready to be channeled to turbines or fields to aid human energy production. Scholars have seen dams as symbols of the modernist project to "tame, control, and discipline nature by converting natural environments into manmade systems.[30]

The Bureau of Reclamation, which runs dams that were built in the twentieth century, stands as one of the most incredible land alteration projects the world has ever seen. The fifty-eight hydroelectric facilities generate fourteen gigawatts of power and irrigate the land on which 75 percent of the nation's vegetables are grown. The most impressive, almost dumbfounding, statistic about the reclamation projects, however, is that they flooded almost sixty-five million acres of land to create reservoirs that allowed for the control of water flowing toward the oceans.[31]

The dams changed the entire geography of the West, affecting where cities could be and where animals could not, what farms made sense and what other activities did not. The Bureau of Reclamation dams are the most sublime examples of the deeper, Promethean truth about energy production in the twentieth century: Creative destruction has ruled. The more complex and astounding the human mechanism, the greater the destruction of natural environments has been necessary to build it.

In fact, it may be the wind and sun's very imperviousness to human control and denaturing that has made them the little-used energy sources. They do not fit into the modern project. At best, they will be a half-tamed naturalized power source. But information technology may be able to flip that disadvantage into a powerful positive. Understanding how to *use* the wind (without controlling it) may both improve power production from renewable energy flows and also provide a new model for how we can live in our human reconstructed world without destroying it.

Human beings don't have a good natural sense for the three-dimensionality and streakiness of air masses moving through the atmosphere. The best we can do to glimpse the chaos and variation is to look up at the individuality of clouds: The same forces that spin wind turbines' massive blades drive all the infinite shapes of the clouds and push them around the world.

Our machines can see much more clearly how the air behaves. Cheap sensors and better algorithms are creating a new type of "environmental awareness" that allows wind engineers to understand the complex natural system they're tapping. Wind turbines, outfitted with laser remote sensing systems, can adjust themselves to the wall of wind approaching their blades. The laser sensors and control systems into which they feed are a sterling example of what the late-twentieth-century technology can do. It's a testament to how far knowledge of the wind has come over the last several hundred years, accelerating in just the last fifty thanks to more measurements and better data crunching.

Humans are excellent but imprecise wind sensors: We can't see it, but we can feel it. Wind has a force, and in keeping with the methods of science at the time, medieval scientists wanted to measure that force. The term "anemometer" (*anemo* means wind in Greek) came into the world in the early eighteenth century, when Peter Daniel Huet, bishop of Avranches in France, designed a strange device to measure the force of the wind, which, sadly, as he related in his autobiography, was never constructed. Huet wrote,

> There had settled at Paris an Englishman named Hubin, a man of ingenuity, and a skillful and industrious workman in mechanics. I went to him and as soon as I mentioned my idea of weighing and measuring the wind, he thought it a matter of jest, and supposed I was ridiculing him. I then produced the figure of a machine by which the force of the wind might easily be weighed as in a balance, and which might be termed an anemometer.[32]

The dominant design for an instrument to measure the wind was sketched out by the Irish astronomer Thomas Robinson in 1846: It's called the cup anemometer and it consists of a few little hemispheres that spin around in the wind. If we attach them to some gears and a counting apparatus, we can get a decently accurate determination of wind speed.[33] Various improvements were made to this data-gathering apparatus over the next hundred years. The cup anemometer is so good now that Leif Kristensen of Denmark's flagship energy laboratory calls them "perhaps even the best" instrument for measuring the main thrust of the wind over time.[34]

Then, in the late 1970s the Department of Energy embarked on an ambitious plan to map out the nation's wind resources. Before that, the best wind spots were more the stuff of local lore than scientific fact. So the Pacific Northwest National Laboratory and the Solar Energy Research Institute began publishing the first regional maps of the country in the early '80s and a national wind atlas in 1986. Ever since, wind analysts have been trying to increase the resolution of those maps in both time and space.[35]

As cheaper computers became available, the company AWS Truewind began marrying large-scale weather models with small-scale topographic maps. They created a parallel process for crunching wind data and ran it on small, fast PCs to get supercomputer-level power at low cost. Then they refined their estimates with data from 1,600 wind measurement towers. Now, high-resolution maps are available that can almost be used to site individual wind turbines.[36]

But that average wind speed data doesn't reflect the second-to-second variations in the wind. Between 1975 and 1985 scientists began to understand that the wind didn't blow the way they thought it did.[37] The wind wasn't like a nice, smooth plane hitting a wind turbine; instead, it had its own structure and it hit like a fist with all those knuckles and bones. Two sites with the same mean wind speed might have very different power production potential because of the individual characteristics of the structure of the wind at each site.

Most wind turbines now come equipped with cup anemometers and wind vanes mounted on their nacelles, the streamlined huts that contain their electronic guts. These anemometers feed data into a computer that controls where the wind turbine faces and the angle at which its blades meet the wind. It's a high-tech variation on the old side vanes that used to keep self-regulating American water-pumping windmills facing into the wind during the nineteenth century.[38]

The anemometers, though, are mounted behind the blades: They know what the wind is like only after it has gone through the business end of the wind machine. Sometimes, the delay between the wind changing and the turbine adapting is a quarter of an hour, which not only sacrifices higher electricity output but can put stress on the out-of-position blades.[39]

That's the problem that Catch the Wind, Inc. is trying to solve with its laser-guidance system for wind turbines: the delightfully evil-sounding Vindicator. A laser mounted to the nacelle sends out pulses that travel up to nine hundred feet, bounce off microscopically small particulates, and then return to a fiber optic detector. The LIDAR system is like sonar for the wind: It allows Catch the Wind's software to create a three-dimensional model of the wind that's flowing toward the turbine. The data then feeds into the turbine's control unit. In an early field trial the LIDAR system increased the production of a wind turbine more than 12 percent.[40] A similar system developed by the Danish National Laboratory for Sustainable Energy's Risø wind research unit generated a 5 percent power increase.[41]

With more than thirty-five gigawatts of wind turbines installed in the United States, a 5 to 12 percent increase in power production would be like adding hundreds of new turbines. Fort Felker, director of the National Renewable Energy Laboratory's Wind Technology Center, thinks that the LIDAR systems may have an even greater impact in reducing operations and maintenance costs for wind farmers. If the LIDAR picks up a freak gust of wind that could damage the turbine, the turbine's controller could take evasive maneuvers to reduce the strain on the machine.[42]

The technology is a nearly perfect marriage of three major strains of R&D in the early twenty-first century: clean energy, aerospace, and telecommunications. Catch the Wind's LIDAR system is an adapted version of a system built for helping helicopter pilots navigate the dusty terrain of Iraq and Afghanistan. It makes economic sense only because the cost of fiber optics and laser diodes were pushed down during the telecom boom of the late twentieth century.[43] All that technology, paid for by the military, is now being used to understand how wind blows so that it can be used to make electricity without using natural gas.[44]

At the same time, a new group of ecologists have begun to argue that the world is so thoroughly dominated by human forces that we're in a new geological era: the Anthropocene, thanks to the introduction of fossil-fuel burning engines two hundred years ago.[45] Pigeons don't live in biomes but rather in anthromes, as do most other animals. We're growing up into a world that can no longer divide itself neatly between human-built and natural worlds. Each is becoming more like the other.

Nature is becoming more human managed, but the human systems that produce energy are becoming less destructive and more closely matched to the environment. Sensor-driven environmental awareness provides the means to make the whole world a human workshop, not just a store of natural resources to be exploited. By using data to understand the world, we can profitably adjust our machines to work with the natural energy flows.

In the 1930s Leo Marx identified the impact of the arrival of industrialization in America in his book *The Machine in the Garden*. The title of his book has been a trope and organizing principle for writing about the American relationship to technology ever since. Seventy years later Adam Rome played on Marx's formulation with a new theme that explained the rise of the modern environmental movement: *The Bulldozer in the Countryside*. Both these images have evoked human mechanical might's dissonance with the land on which it works.

Our times are stranger. Humans are refiguring their relationship with natural systems to both control more and destroy less at a time when local areas are increasingly part of global networks. We now create both the mechanical and natural worlds, but fully control neither. Perhaps it's time for a new vision of the American relationship between technology and environment: the Turbine in the Anthrome.

chapter 27

Rehumanizing Environmentalism

HERE ARE TWENTY-FIVE or so desert tortoises crawling around a four thousand–acre patch of the Mojave Desert known as the Ivanpah Valley. A minor biological marvel, these reptiles are able to survive in temperatures of up to 140 degrees and go for a year without access to water. About a foot long, and maybe a dozen pounds, they don't look like much.[1] But this tiny little band of creatures, and others like it, may be the key lever that environmental groups use to prevent large-scale solar installations from blossoming in the vastness of California's arid lands.

If ever there is going to be a place where solar energy works, the Mojave Desert is it. It is, as venture capitalist and solar enthusiast Bill Gross likes to say, the Saudi Arabia of solar. Even better, it's close to the large electricity markets in southern California and Las Vegas. Looking at a map of America that's rainbow color–coded for the most photons of sunlight available to be turned into electrons, the Mojave glows like a big red button that says to green-tech entrepreneurs, "Push here!"

But it so happens that of the 16 million acres of the Mojave Desert, 4.6 million of them are considered to be "critical habitat" for the tortoise. Putting a solar plant anywhere in—or near—that habitat requires extensive off-setting measures, if the plant can even be built despite preservationist opposition.

To make matters worse, the desert tortoise and solar developers have the same good taste in Mojave terrain. Both like "nice, broad valleys" that are relatively flat and receive huge amounts of solar energy.

It's more than a theoretical issue. The presence of the handful of desert tortoises per square mile of land has been a huge issue for BrightSource, the descendent of Luz International, as they attempt to build a solar plant in the Ivanpah. The solar thermal power plant uses fields of mirrors to redirect the heat of the sun onto a boiler, which generates steam that drives a turbine. It's a fairly well-established technology that can be deployed at the same size as fossil-fuel plants. At 400 megawatts of capacity, the plant would be like 100,000 or more average home solar–PV installations: That's 40 percent more arrays than all of the solar panel installations ever put in by Americans.[2] Ivanpah alone would nearly double the solar capacity in California, a state that's told itself that it must receive 33 percent of its electricity from renewable sources by 2020. It's nearly impossible to imagine a scenario in which California is able to do that without large-scale solar power plants in the desert. Dozens of solar thermal companies are lining up to make sure that the state doesn't have to.

Societies' deployments of technology have surprised before, but it seems very likely that if California is going to put in thirty gigawatts of renewable energy, some big chunk of it is likely to be in the desert. Under the biggest deployment scenarios, something like forty thousand acres of Mojave may be developed for solar power in coming decades.[3]

The desert tortoise could quickly become the spotted owl of the solar energy industry. It is the creature that has both symbolic power for environmentalists who have been dedicated to its preservation for decades and statuary protections under the Endangered Species Act. The tortoises can act as a legal lever to protect whole swaths of the Mojave Desert from green-tech development, serving "as vulnerable symbols of biological diversity while at the same time standing as surrogates for wilderness itself," in the words of historian Bill Cronon. The form of the laws has forced environmental groups to use single species in this way, "thereby making the full power of the sacred land inhere in a single numinous organism whose habitat then becomes the object of intense debate about appropriate management and use."[4]

And that is the situation shaping up in the Mojave. Except in this case, it's a bit unusual. Both sides can legitimately claim the mantle of protecting the environment. One side protects *this patch* of wilderness, whereas the other protects a dispersed patch of atmosphere from carbon dioxide emissions, an invisible substance present in tiny concentrations. It's telling that no power plant ever inspired more organized groups to comment on a plant than Ivanpah, not even the Sundesert nuclear plant, the symbolic end of nuclear power in California.[5]

Ivanpah is a bellweather, then, and environmental groups in California, battle-hardened by years of fighting power plants, were quick to organize to critique the project and position themselves for a protracted public and legal struggle. The arena for this intra-green battle is the California Energy Commission's power plant siting process, presided over by Commissioner Jeffrey Byron. At the opening of the near-final round of hearings in January 2010, Byron said,

> I want to make clear to everybody that this is an extremely important project for this Commission . . . this represents the first of what we hope will be many renewable projects that will come before the Energy Commission. It's unique in that regard, but also in that it has a number of large land use issues associated with it, and biological and others that will come up.[6]

Outside groups, known in the jargon of the process as "intervenors," have multiple opportunities to submit legal briefs, provide testimony, and cross-examine whomever they would like from the project developer's personnel bench.

In hearings before the Commission, BrightSource officials and contractors have had to answer for absolutely every detail of the project's plan, from the distance of the mirrors to the tower boiler that receives reflected solar heat to the type of materials that can be used in the fences that will surround the site to keep out stray tortoises. Their biologists and engineers are subjected to blistering questioning. Environmental groups have called thousands of the tiniest details into question. Intervenors can and have proposed radically different alternative plans, sites, and means of generation, long after BrightSource had purchased this piece of land from the Bureau of Land Management.[7]

As nuclear advocates have complained for forty years, whether the environmental intervenors want to modify specific proposals or protect certain species or whether they just want to shut down the whole damn thing and prevent it from being built is not always clear. In press releases and the pages of the country's major newspapers, environmental groups jockeyed for position. Some, like the Sierra Club, tried to chart a middle path. "It's not enough to say no to things anymore," said one of the group's experts on renewable energy. "We have to say yes to the right thing."[8]

But it's clearly an uncomfortable position. When BrightSource altered its plan to reduce its output and change its design in response to criticisms, the Sierra Club responded with a mixed statement. "Looking at this new proposal, it will not do anything to protect the desert tortoise and they won't be able to generate as many megawatts," said the group's senior attorney in San Francisco, Gloria D. Smith. Despite that, she said, "We still support this project but just want it to have a more beneficial footprint."

The National Resources Defense Council has taken much the same pragmatic stance, with its longtime lawyer Johanna Wald pointing out what is the great irony of the parable of the tortoise and the sun. "We have to accept our responsibility that something that we have been advocating for decades is about to happen," Wald said. The days of a fossil fuel–powered grid may be coming to an end, but new problems are creeping in.

Other groups, particularly the Center for Biological Diversity and Defenders of Wildlife staked out more hardline stances. Kim Delfino wrote a warning in the group's magazine that "California is starting to see a new kind of 'gold rush', but this time, it's going to be our wind, sunlight and public lands that are up for grabs."[9] Large-scale solar farming is, to Delfino, another form of mining and quite possibly a government giveaway to private companies. Defenders of Wildlife harangued the Ivanpah project tooth and nail through the plant-siting process.

One can almost imagine the fossil-fuel industry laughing all the way to the slag pile. Would-be nuclear plant builders, too, must be enjoying watching the solar boys getting the same workout that has kept California from building large coal and nuclear plants. Using primarily solar fuel, or as BrightSource chairman Arnold Goldman calls it, "fuel for life," has not caused environmental groups to give them a free, or even an easy, pass.

For fledgling green-tech companies, the lack of support from those who would seem to be their natural allies could prove to be their Achilles' heel. After decades of fighting power plants and pollution, trying to impose limits on society's activities, mainline conservation groups do not find supporting the destruction of desert by a private power plant developer easy. Like a longtime opposition political party suddenly handed the keys to the kingdom, the environmental movement is discovering that governing is a lot harder than it looks; the cracks in the coalition are easier to see in the realm of action.

In the condensed narrative, the environmental movement rose up in affluent, suburban communities after World War II to oppose environmental pollution and toxic chemicals, which were killing off animal species and shortening human lives. Combining older conservationist thought with new scientific understanding of ecology, environmentalism offered a political alternative to the bipolar Cold War world, critiquing the nasty environmental records of the Soviets and Americans.

Rachel Carson's 1962 book *Silent Spring* is often given credit for launching popular interest in environmental issues.[10] In elite circles she and other ecologists brought new scientific expertise to bear on how nature worked, countering the expertise of other groups of established authorities like utility executives and corporate industrial chemists. "The development of atomic energy, the chemical revolution in agriculture, the proliferation of synthetic materials, and the increased scale of power generation and resource extraction" gave the '60s generation many new technologies that all seemed unabashedly and fundamentally unnatural.[11] Reacting against this tremendous assault on nature, twenty million Americans rose up on April 1, 1970 for the first Earth Day and then scared the Nixon administration into signing the greatest streak of environmental legislation in American history.

Rice Odell's book for the Conservation Foundation on the occasion of the 1980 Earth Day was clear on exactly how big the change was. Odell wrote,

The environmental awakening was a rude one, indeed. It cast doubt on some of the most cherished credos of the day—beliefs in the ever-increasing rewards of industrial, technological, and chemical progress and of economic growth under a free-market system. It made the future look bleak and disappointing. Paradoxically, the Environmental Revolution also brought a glow of welcome reform, of hopeful change, and of prospects for a higher quality of life. It brought a new dimension to civilization.[12]

The primary concerns of the movement's members were protecting wilderness and disabusing Americans of the desire to think of all of nature as a "resource" just awaiting human exploitation. As Carson put it, "The 'control of nature' is a phrase conceived in arrogance, born of the Neanderthal age of biology and philosophy, when it was supposed that nature exists for the convenience of man."[13]

Environmental protection was not simply a matter of keeping a few species alive; rather, it was a necessary component to "saving the earth." Antiwar sentiment also ran through these sentiments. "Peace, as has often been said, is indivisible," E. F. Schumacher wrote. "How then could peace be built on a foundation of reckless science and violent technology?"[14] Practically speaking, antiwar protest networks often converted to other forms of activism.[15] They brought their opposition to anything that was associated with what Thomas Hughes has called the military-industrial-university complex.

Ken Smith of the Seattle-based environmental community Ecotope, stated in 1973 that "Since the 1960s, a vision of energy conservation has grown. The vast resources wasted in the Vietnam War while millions starved explicated the problem. Now we have moved from antiwar to environmental activism."[16] Smith and many activists from protest movements saw problems with the environment as merely one strand of a society that was dangerously out of whack.

George McGovern, in his introduction to *Defending the Environment*, a book by environmental lawyer Joseph Sax, wrote, "In the United States today frustration abounds in nearly every area of human concern—the ending of a tragic war, the eradication of the silent violence of hunger, the extension of racial and human justice, and the reversing of an inexorable environmental tragedy which began two centuries ago."[17]

What's more, the ongoing military buildup of the Cold War obviously exacerbated this general unease with the way things were going. Nuclear weapons provided humanity with its first real opportunity to destroy the entire world, which would clearly have been quite an environmental issue. Leaders like Paul Ehrlich, a Stanford biologist, were quick to connect nuclear weapons with what was often termed a war on the environment. Near-term survival of humans as a species and the earth as a host for life was genuinely considered to be questionable. Faith in humanity to avert near-term catastrophe ran low. Few would have anticipated the go-go late '90s.

Ehrlich's book, *Population Bomb*, illustrates the tangle of these ideas. One of the most popular books of the late '60s and early '70s, the *Bomb* went through twenty-two printings between May of 1968 and November of 1970 alone. Ehrlich is remembered largely as a "neo-Malthusian" who argued that a vast human die-off in the developing world was going to occur because the world would run out of food.

But this shorthand bottling of his ideas shortchanges the apocalyptic strangeness that one gets reading the book now. *The Population Bomb* intersperses population dynamics research, policy talk, a zero population–growth manifesto, and very bad short fiction. It shows how tightly bound nuclear and environmental fears were: "The End is Near" was always near. Ehrlich's nuclear holocaust scenario went:

> Freddy was happy behind the plow. The mule was strong, and the work was going well. Only in some parts of the southeast had survival been possible and he'd been one of the lucky ones. His mother had died after a twenty-year battle with radiation-induced illness; but they'd had some good times, and she's lived to see him marry Louise. If only she'd lived to see the baby born and known that it was all right. So few were.[18]

Other characters in that scenario die or contemplate suicide with cyanide. This is as dark a vision of the future as it is possible to imagine—an eco-horrorshow. And this was what a Stanford professor was writing for a popular audience! In Ehrlich's own words, the most optimistic of his three scenarios "has considerably more appeal than the others, even though it presumes the death by starvation of as many as a billion people. . . . I challenge you to create one more optimistic."[19] (Luckily, we have all lived one that is many times over more optimistic.)

Guided by a new knowledge of ecological systems, a faith that humans should live "in harmony with nature," and a fear that the world might end soon, environmental interest groups became deeply entrenched in rich countries the world over.[20] Harvard sociologist David John Frank argues,

> There has been a shift away from conceptions of nature as a realm of chaos and savagery, and away from conceptions of nature as a cornucopia of resources, and toward conceptions of nature as a universal, life-sustaining "environment" or ecosystem. The new conception is that of a natural system with planet-wide interdependencies, encompassing Homo sapiens and providing most fundamental sustenance for this species. This redefinition of nature, championed by scientists and environmental activists, facilitated the striking level of world mobilization that has emerged.[21]

By the 1970s a common belief system emerged. Environmentalists would have agreed with statements like "We are severely abusing the environment" and "There are limits to growth beyond which our industrialized society cannot expand." They would have disagreed with the idea that "humans have the right to modify the natural environment for our own ends."[22] In the energy world, the environmental movement promoted "soft" technologies like wind and solar. Hard technologies like hydroelectric facilities as well as coal and nuclear plants were anathema. And, of course, environmentalists were identified with hippies.

This story of the environmental movement is tidy and coherent, and it possesses a good kernel of truth. But like most creation stories, it hides some things and emphasizes others. The raw material of what happened has been pounded into new and useful forms by political groups and their enemies. "Even though we use the phrase 'the environmental movement,' it's bullshit," said Adam Rome, a Penn State professor who is probably the world's scholarly authority on the genesis and origins of environmentalism in America.[23] It's simply not an accurate depiction of the incredible variety of people who wanted to reconfigure our relationship with the nonhuman world. Not all of them were primarily interested in protecting biodiversity, and many of them were just fine with modifying the natural environment for human ends. Historian Andrew Kirk wrote,

> One of the popular misconceptions about environmental advocacy in American history stems from the desire to celebrate the few individuals who advocated the preservation of nature where humans weren't, while often ignoring those who worked to use their technological enthusiasm to benefit nature. Historical actors in the drama of twentieth-century environmental advocacy are often rated on a sliding scale according to the purity of their wilderness vision.[24]

There is an alternate vision, though, that this book has tried to highlight and that historians like Rome and Kirk have begun to excavate. From the late 1950s onward, traditional Democratic liberals—the FDR type, not the eco type—had a pretty coherent program for making the country better: Boost public spending on the social goods that private enterprise seemed to neglect, including environmental protection to provide "qualitatively" better lives for a large middle class that had it all. The impulse would eventually underpin Lyndon Johnson's "Great Society" programs as well as Nixon's environmental program.

The most-quoted passage from the economist John Kenneth Galbraith's 1958 best-selling book *The Affluent Society* described the sorts of problems that American society began to encounter in the years after World War II. In it, increasingly nice private goods shuttle and protect Americans from a deteriorating public environment. Galbraith thundered,

> The family which takes its mauve and cerise, air-conditioned, power-steered, and power-braked automobile out for a tour passes through cities that are badly paved, made hideous by litter, blighted buildings, billboards, and posts for wires. . . . They picnic on exquisitely packaged food from a portable icebox by a polluted stream and go on to spend the night at a park which is a menace to public health and morals. Just before dozing off on

an air mattress, beneath a nylon tent, amid the stench of decaying refuse, they may reflect vaguely on the curious unevenness of their blessings. Is this, indeed, the American genius?[25]

The strength and ductility of the environmental movement came through little environmental injustices like the ones Galbraith described, encountered in the new, treeless suburbs and the streets of the cities. Different subsets of Americans—often led by women like iconoclastic economist Hazel Henderson—began to coalesce around the idea that clean air and water were worth paying for.[26]

Note that this environmentalism is not nature, endangered species, or wilderness focused: It is concerned, first and foremost, with humans. Clean and safe were more important than "natural." The suburban housewives and baseball dads who supported the passage of the nation's landmark environmental legislation were not interested in biomes, per se: They cared about the places that humans co-created, or, as ecologist Erle Ellis calls them, anthromes.

This tension between different environmental thinkers is a live issue. One exchange in the BrightSource hearings was particularly telling. Michael Connor of the Western Watersheds Act was cross-examining an ecologist hired by BrightSource about the possibility of ever restoring the land that would be used by the plant. Connor asked W. Geoffrey Spaulding, who had thirty-five years of botanical experience in the Mojave, to explain how the restoration of an area of the desert disturbed by the construction of a pipeline had worked.

"Re-vegetation actually, in comparative terms, over the last nine or ten years, seems to be going along fairly well," Spaulding testified.

"And so you think it looks—it's starting to look natural?" Connor asked.

"Please define natural," Spaulding fired back. The restoration was actually the height of unnatural efforts made by human beings, first to disturb the land and then to make extraordinary and long-term efforts to restore it.[27]

The confusion resulted because Connor had substituted the word *natural* when what he really meant was *good*. It's a common thing to do in many environmental circles, one that reflects the particular late '60s and early '70s sensibility that the pollster Daniel Yankelovich called "the new naturalism."[28] As related by Rome, "Yankelovich discovered a widespread conviction" among college students that "everything artificial was bad, while everything 'natural' was good." Defining natural, unfortunately, is difficult.[29] In a world completely dominated by humans, what's natural?

THINK GLOBALLY, DESTROY LOCALLY

Big-thinking climate scientists say that solutions to global warming need to meet a key criterion: They must be able to scale up to the size of the problem. This simple dictate will require major changes in the way environmentalists shaped by this era think.

"Think Globally, Act Locally," became a mantra for '70s-trained environmentalists. Stopping your local power plant was a way of protecting the earth. When environmentalism was perceived as a way of reconnecting with the land, then it made sense to work locally, where one could go out and smell the fresh air one was protecting. Large-scale problems like "the military-industrial complex" could be attacked at the grassroots level through many different organizations. Besides, privileging the local ecosystem over the globalized world economy was easy. But what if responding to climate change with renewable energy deployments means sacrificing local landscapes?

That's the very situation that BrightSource's Ivanpah plant presents to nature preservation groups like Defenders of Wildlife. On the one hand, if no mirrors were installed in their territory, this would obviously be better for the desert tortoises. On the other hand, the plant is a key test case for an entire field of technology that could provide large amounts of low-carbon electricity. If Ivanpah works, it could pave the way for dozens of similar projects across the world.

Fledgling energy companies want—and need—to get big if they want to compete with oil, coal, and gas. Scale has a very important meaning for technologists, too. Getting big and making a lot of something are how companies drive down the unit cost of a product. It's a lesson as old as Henry Ford and as broad as Walmart. As solar investor Bill Gross likes to say, the only way to get the price of something down near its material cost is to make a million of that something. It's true for cars and it's true for the mirrors that power these fields.

Yet it's exactly that scenario that is most frightening to wildlife environmentalists.

"I'm very concerned about the whole scale of this. I think you really need to take a strong comprehensive look at what's going on here," one desert tortoise biologist testified before the Commission. "It's a very complex situation, and there's a lot of actors involved, and it really needs a wide sweeping programmatic look."[30]

If there is one problem with the solar power stations in the desert, it's scale—sheer size. Although the impacts of one or two projects could probably be mitigated, a push to generate considerable amounts of our electricity from solar without dramatically cheaper solar photovoltaic panels basically requires building solar farms in the desert. The California Energy Commission explicitly considered whether rooftop solar could meet the state's renewable energy generation goals, and the answer was a resounding no.[31] The environmental movement, as currently constituted, will have to learn how to deal with that build out. Coming up with ways of reconciling the need for low-carbon energy with the desire to protect endangered species and wild habitat has to be the dominant intellectual challenge for greens of the next generation.

This is because the battles in Sacramento aren't just about California. The tension on display in the hearing rooms of the California Energy Commission runs straight back to the heart of environmentalism and its complex relationship with technology. Sociologist William Gamson argued that Americans have long believed in "progress through technology," but wrapping around and cutting through this dominant framework is a countertheme that humans should live in "harmony with nature."[32] The Jetsons epitomize the first idea whereas commune-living hippies embody the second. If there's a big, definable arc to the American relationship with technology, this is it.

What's important about green technology is that it may resolve a tension that's threaded through history from John Etzler and Henry David Thoreau, through the meatpacking plants of Chicago, past the oil fields of Texas, beyond the solar homes of New Mexico, to the suburbs of Los Angeles, and up to the Trans-Alaska oil pipeline. For environmental groups, the answer won't be coal power or no power plant at all. There will be real alternatives that can be promoted and supported.

At the same time that our technology is coming to depend on natural energy flows, scientists are discovering that "nature" is a lot more manufactured than it looks. Ellis and others point out that humans have altered nearly the entire surface of the globe in ways that are at least as deep as what we're doing to the atmosphere by burning fossil fuels. Ellis wrote,

> Humans have transformed more than three quarters of the terrestrial biosphere into croplands, rangelands, villages, settlements and other anthropogenic biomes (anthromes) including managed and recovering woodlands. While the process of anthropogenic ecosystem transformation and management has been sustained for thousands of years on every continent except Antarctica, there is still a tendency for ecological scientists, educators and policymakers to portray the terrestrial biosphere as a natural place just recently disturbed by humans.[33]

The hundreds of pages worth of ecological testimony by the environmental intervenors arguing over methodological differences in counting desert tortoises seem to lack perspective. The solar plant's human components were put on trial by groups and institutions that developed in order to stop the building of plants, not support them. The lives of a few dozen desert tortoises may very well improve as a result of the process, but the opposition to Ivanpah raises questions about how unsympathetic some environmental groups are willing to be about the realities of running a solar firm that can compete with fossil fuels.

Despite the continuing opposition, in March of 2010 the California Energy Commission recommended that the Ivanpah plant move forward. "The Ivanpah Solar Electric Generating System (ISEGS) project is iconic of the coming transformation of the electric generation system in California, and perhaps the country as a whole," the CEC's staff explained.[34] Although BrightSource had to make considerable concessions, they will be able to build their plant. That is a good thing. Brightsource broke ground in October 2010 at a star-studded event featuring California governor Arnold Schwarzenegger and the Secretary of the Interior Ken Salazar. "Some people look out into the desert and see miles and miles of emptiness," Schwarzenegger said. "I see miles and miles of gold mine."[35]

There is no turning back on the enormity of human civilization's impact on the globe. Now is the time to recognize that even the wildest Amazonian and Mayan jungles are feral landscapes that have been permanently and massively altered. "We've got to stop trying to save the planet," Ellis wrote in a WIRED Science article. "For better or for worse, nature has long been what we have made it, and what we will make it."[36]

Viewed as an anthrome, as a human space, the Mojave already has seen massive alterations. For one, there are already nine solar electric generating stations built by BrightSource's predecessor Luz International. A couple of them cover 250 acres of land near Kramer Junction. Most traditional accounts of the solar plants emphasize how massive the plants are. "The rays captured on the huge, rain-gutter-shaped mirrors fire sleek tubes of synthetic oil, which in turn generate enough turbine-driving steam to power more than 100,000 homes," wrote Paul Pringle for the Dallas Morning News in 1989. One expects them to dominate the desert landscape, but in context, there was little dissonance between the plants and their surroundings. They were not out of scale.

Maybe it's because just down the road, one of the world's largest boron mines has made a hole in the earth that's bigger than the mirror field and five hundred feet deep. Maybe it's because on the way to Kramer Junction we drive past the giant logistics companies near Victorville, with their million-square-foot warehouses connected to the entire world by plane and train, the road-facing flanks of the buildings perforated with hundreds of semi-bays for trucking plastic toys and lawn furniture the last mile of the journey. Maybe it's because on the drive back to the coast, we pass through the haze-filled San Joaquin Valley, in which a few mechanically enhanced hired hands create food on a truly impressive scale: miles and miles of fields feed just a tiny slice of our hunger for almonds. The whole enterprise is tawdry and sublime at the same time.

Out there, in the Mojave's industrial context, where lots of things are happening on the scale of the global economy, having 250 or 2,500 acres of mirrors making steam to turn a turbine is not that weird. Nature-loving city people don't go to the human areas of the Mojave for a reason: That country is big, mechanical, and fast. It is the opposite of what people wander into the wilderness for. But that's where these solar plants are going, out where the land is cheap, where energy and stuff are made and moved. The Ivanpah solar power plants may seem huge, but they will eventually take up only 0.025 percent of the Mojave. In the geography of infrastructure and on the scale of the global climate, a million square meters of mirrors may be defined as small and beautiful—or at least sublime.

People conceive of the American wilderness, Bill Cronon argued, as "the ultimate landscape of authenticity. Combining the sacred grandeur of the sublime with the primitive simplicity of the frontier, it is the place where we can see the world as it really is, and so know ourselves as we really are—or ought to be." But this is a problem because "the dream of an unworked natural landscape is very much the fantasy of people who have never themselves had to work the land to make a living."[37] The only way that cities and wilderness exist as they are is because of all those other things we stick out in the Mojave. Energy, industrial, and commercial facilities are the lifestyle-support system of our country. It is in the infrastructural landscapes we've scratched into far-flung natures where we can see actual human society reflected.

What better symbol could there be of who we really are—or ought to be—than a field of mirrors harnessing the sun to make huge amounts of electricity in the middle of nowhere, surrounded by a fence to keep out desert tortoises who have had their homes moved to a carefully constructed new location as if they were very high-paid executives switching jobs? That's living with our world—and it's what a naturalized, if not natural, energy system looks like, the kind humans could live with for a very long time.

Green technology gives environmentalism the material means to build a better civilization as well as the political potency and clarity of purpose that comes with the need to make new things work.

NOTES

INTRODUCTION

1. Historic American Engineering Record, "Death Valley Ranch."
2. Edgerton, *The Shock of the Old*, xvii.
3. Ibid., 8.
4. Shapiro and Varian, "The Art of Standard Wars."
5. Campbell-Kelly, "Not Only Microsoft."
6. Economides, "Competition and Vertical Integration."
7. Hughes, "Technological Momentum in History."
8. Winner, *The Whale and the Reactor*, 80.

PART I:
THE DREAM OF A MORE PERFECT POWER

CHAPTER 1

1. "John Doerr Sees Salvation and Profit in Greentech."
2. Kedrosky, "John Doerr's Tears."
3. Smil, "Moore's Curse."
4. LaMonica, "Green-tech Venture Investing."
5. Garber, "John Doerr."
6. "Bill Gates on Energy."
7. Garber, "John Doerr."
8. Senate Committee on Public Works, *The Impact of Growth on the Environment.*
9. Thiele, *Environmentalism for a New Millennium*, 55.
10. Stoll, *The Great Delusion.*

CHAPTER 2

1. Etzler, *The Paradise Within the Reach of All Men*, 87.
2. Stoll, *The Great Delusion*, 47.
3. Noyes, *History of American Socialisms.*

4. Williamson, "Urban Disamenities."
5. Steckel, "Stature and Living Standards."
6. Brimblecombe, "London Air Pollution."
7. Mrozowski et al., "Living on the Boott."
8. Dickens, *Oliver Twist*, 29.
9. On Etzler, see Stoll, *The Great Delusion*, 38.
10. Noyes, *History of American Socialisms*, 19.
11. Stoll, *The Great Delusion*, 53.

CHAPTER 3

1. A. MacDonald, "Lowell," 37.
2. Nye, *American Technological Sublime*, 39.
3. Mackay, *The Western World*, 282.
4. Vigne, *Six Months in America*, 236.
5. Whittier, "The Stranger in Lowell."
6. Dalzell, *Enterprising Elite*.
7. Nye, *American Technological Sublime*.
8. Dalzell, *Enterprising Elite*.
9. "Editorial: Letter IX," 449.
10. Case, "Notes from a Journal," 181.
11. "Editorial: Letter IX," 449.
12. Lyford, *The Western Address Directory*, 37.
13. Montrie, "'I Think Less of the Factory,'" 284.
14. Wyman, "Practical Treatise on Ventilation," 132.
15. Ibid., 133.
16. Lyford, *The Western Address Directory*, 88.
17. Crockett, *An Account of Col. Crockett's Tour*, 95.
18. Brierley, *Ab-o'th'-Yate in Yankeeland*.

CHAPTER 4

1. Steinberg, *Nature Incorporated*.
2. Cumbler, "Review of *Nature Incorporated*," 626.
3. Steinberg, *Nature Incorporated*, 69–76.
4. Ibid., 49, 79–80, 167–97.
5. Hunter, *Waterpower*.
6. Thoreau, *A Week on the Concord and Merrimack Rivers*, 40.
7. Malone, *Waterpower in Lowell*.
8. Steinberg, *Nature Incorporated*.

9. Thoreau, *A Week on the Concord and Merrimack Rivers*, 44.

10. Thoreau, "Paradise (to Be) Regained," 452.

11. Linstromberg and Ballowe, "Thoreau and Etzler," 25.

12. Thoreau, "Paradise (to Be) Regained," 459.

13. Ibid., 460.

14. Stoller, "A Note on Thoreau's Place," 173.

15. Miller-Rushing, "Global Warming and Flowering Times"; Willis, et al., "Phylogenetic Patterns of Species Loss."

16. Willis, et al., "Phylogenetic Patterns of Species Loss," 17031.

17. Brand, *Whole Earth Discipline*, 1.

18. "Secretary Chu Announcement."

19. Reich, *The Greening of America*, 350.

20. Obama, "Remarks by the President."

21. "Earth's Boundaries?"

22. Winner, *The Whale and the Reactor*, 9.

PART II: WHAT WAS

CHAPTER 5

1. Nye, *Consuming Power*.

2. Rosenberg and Trajtenberg, "A General-Purpose Technology at Work."

3. Nye, *Consuming Power*, 72–73.

4. Rosenberg and Trajtenberg, "A General-Purpose Technology at Work," 62.

5. Nye, *Consuming Power*.

CHAPTER 6

1. Webb, *The Great Plains*, 346.

2. Righter, *Wind Energy in America*, 26.

3. Torrey, *Wind Catchers*, 103.

4. Webb, *The Great Plains*, 341.

5. Murphy, *The Windmill*.

6. Rosenberg and Trajtenberg, "A General-Purpose Technology at Work."

7. Righter, *Wind Energy in America*, 25.

8. Perry, *Experiments with Windmills*.

9. Edwards and Edwards, *Batavia*.

10. Perry, *Experiments with Windmills*, 20–30.

11. Ibid., 38–44.

12. See, for example, Walker, "Reliability and the Future of Wind Energy."

13. Baker, *Field Guide to American Windmills.*
14. Perry, *Experiments with Windmills*, 23–25.
15. Gipe, *Wind Power*, 248.
16. Smeaton, *Experimental Enquiry.*
17. Perry, *Experiments with Windmills.*
18. Baker, *Field Guide to American Windmills.*
19. Ibid., 36.
20. Baker, *Field Guide to American Windmills*, 38.
21. Kutleb, "Can Forests Bring Rain?"
22. Smith, "Rain Follows the Plow."
23. Aughey, *Sketches of the Physical Geography*, 44.
24. Wilber, *The Great Valleys and Prairies*, 73.
25. Hansen and Libecap, "Small Farms, Externalities."
26. A. N. Williams, *The Water and the Power*, 339.
27. Stegner, *The American West as Living Space.*
28. A. N. Willams, *The Water and The Power*, 339.
29. Boyer, et al., *The Enduring Vision*, 607.
30. Smith, "Rain Follows the Plow."
31. Webb, *History as High Adventure*, 37.
32. Barbour, *Report of the State Geologist.*
33. Flint, *American Farming and Stock Raising*, 537.
34. Barbour, *Report of the State Geologist*, 34.
35. Knopp, "Mammoth Bones," 189.
36. Barbour, *Report of the State Geologist*, 37.
37. Murphy, *The Windmill*, 138.
38. Barbour, *Report of the State Geologist*, 30–35.

CHAPTER 7

1. Black, *Petrolia.*
2. Ibid., 20.
3. Ibid., 21.
4. Eaton, *Petroleum*, 214.
5. Yergin, *The Prize*, 11–12.
6. Black, *Petrolia*, 18.
7. *Robert Dennis Collection.*
8. Black, *Petrolia*, 112–114.
9. Yergin, *The Prize*, 12.
10. Stein, *When Technology Fails*, 30.
11. Black, *Petrolia.*

CHAPTER 8

1. Levy, *920 O'Farrell Street*, 185.
2. Wolfe, *Rube Goldberg*.
3. *Prospectus for the Wave Power and Air Compressing Company*.
4. Duffy, "Wave-Power Air-Compressor."
5. *Prospectus for the Wave Power and Air Compressing Company*, 2.
6. Wave Motor Patent Database.
7. A common dictionary definition. See, for example,
 www.yourdictionary.com/rube-goldberg.
8. Wolfe, *Rube Goldberg*, 38.
9. Duffy, "Wave-Power Air-Compressor," 1.
10. "Destitute of Meaning," Feb. 27, 1898, 4.
11. "A Wife's Appeal."
12. *Prospectus for the Wave Power and Air Compressing
 Company*, 2.
13. "The Best Idea of All."
14. "Notes by a Roving Contributor—13," 367.
15. "'Edison Said Only a Few Years Since' [advertisement]."
16. Bennett, "Wave-Motors," 251.
17. Melville,"The Engineer and the Problem," 820.
18. "Reynolds Wave Motor and the Inventors."
19. Ibid.
20. *California Wave Motor Prospectus*.
21. Reynolds,"Pile Cleaner and Protector."
22. Wallace, "Unlimited Electric Power," 289.
23. "A Smokeless City."
24. "Wave Power Chiefs Quit."
25. "Redondo Pier Sinks in Sea."
26. *The San Francisco Call*, March 28, 1909.
27. *Prospectus for the Wave Power and Air Compressing Company*, 3.
28. Dalton, Alcorn, and Lewis, "Case Study Feasibility Analysis,"
 443–455.
29. S. Williams, "U.K. Tries to Catch a Wave."
30. *California Wave Motor Prospectus*, 1.

CHAPTER 9

1. J. C. Williams, *Energy and the Making of Modern California*, 82–86.
2. E. D. Adams, *Niagara Power*, 167.

3. "Pneumatic Clocks," *New York Times.*

4. "Pneumatic Clocks," *Scientific American*, 19.

5. Cole, "Underground Social Capital."

6. Steenson, "A Regressive Glossary of the Poste Pneumatique."

7. Kennedy, *Compressed Air*, 50.

8. Ayers, *The Architecture of Paris*, 208.

9. Saunders, *Compressed Air Information*, 220.

10. Calculated from Ayres and Warr, "Dematerialization vs. Growth."

11. Kennedy, *Compressed Air*, 51.

12. Foran, "The Day They Turned the Falls On."

13. E. D. Adams, *Niagara Power*, ix.

14. Ibid., 423.

15. Foran, "The Day They Turned the Falls On."

16. Ryan, *High Voltage Engineering and Testing*, 331.

17. EIA, "EIA State Energy Profiles."

18. Schatzberg, "The Mechanization of Urban Transit," 226–229.

19. Sanderson and McGovern, *The World's History and Its Makers*, 235–236.

20. "The Compressed-Air Motor."

21. "New Power for Elevated."

22. Nye, *Electrifying America.*

23. J. C. Williams, *Energy and the Making of Modern California*, 176.

24. *San Francisco Call.*

25. U.S. Department of Energy. "GridWorks."

PART III:
WHAT MIGHT HAVE BEEN

CHAPTER 10

1. Abernathy, "Progressivism," 524.

2. Salom,"Automobile Vehicles," 281.

3. Mom, *The Electric Vehicle*, 62.

4. Salom,"Automobile Vehicles."

5. Ibid., 278.

6. Ibid., 289.

7. Shulman, Personal communication.

8. Appleton,"Recent Developments in the Applications of Storage Batteries," 185.

9. Rae,"The Electric Vehicle Company."

10. D. Kirsch, *The Electric Vehicle and the Burden of History*.

11. Appleton,"Recent Developments in the Applications of Storage Batteries," 184.

12. "Mr. William C. Whitney's New York House."

13. Morris and Salom, "Letter to Western Society of Engineers," 547.

14. D. Kirsch, *The Electric Vehicle and the Burden of History*, 42.

15. Ibid., 42–43.

16. Ibid., 54.

17. Salom,"Automobile Vehicles."

18. Engerman, *Cambridge Economic History of the United States*, 629.

19. Duncan,"The Future of Electricity in Railroad Work," 410.

20. Rae,"The Electric Vehicle Company."

21. Aronson,"The Sociology of the Bicycle," 309.

22. Norcliffe, "Popeism and Fordism," 270.

23. Ibid., 267.

24. Pope,"Export of American Bicycles," 181.

25. Epperson, "Failed Colossus," 316–318.

26. Pope,"Export of American Bicycles," 181.

27. Bellamy, *Looking Backward*, 75–76.

28. Ibid., 78.

29. D. Kirsch, *The Electric Vehicle and the Burden of History*, 61–73.

30. Rae,"The Electric Vehicle Company," 302–309.

31. D. Kirsch, *The Electric Vehicle and the Burden of History*, 79.

32. Aronson,"The Sociology of the Bicycle."

33. Hugill,"Good Roads and the Automobile," 328.

34. Aronson,"The Sociology of the Bicycle," 309.

35. Ibid., 310.

36. Hugill,"Good Roads and the Automobile," 328.

37. Aronson,"The Sociology of the Bicycle," 311.

38. McCarthy, *Auto Mania*, 12.

39. F. L. Allen, *The Big Change*.

40. "Whitney Auto Smash."

41. Kirsch and Mom, "Visions of Transportation," 78.

42. McCarthy, *Auto Mania*.

43. "Who Owns It, Anyway?"

CHAPTER 11

1. National Park Service, "Historic Structure Report."

2. National Park Service, "Scotty's Castle—Behind the Scenes."

3. Butti and Perlin, *A Golden Thread*, 140–141.
4. "Growth of Idea Is Told by Inventor."
5. Ibid.
6. Butti and Perlin, *A Golden Thread*, 129.
7. Bailey,"Solar Heater."
8. Butti and Perlin, *A Golden Thread*, 120–121, 123–127.
9. Ibid., 121.
10. Ibid., 117–118.
11. Ibid., 132–133.
12. J. C. Williams, *Energy and the Making of Modern California*, 36.
13. Ibid., 130–137.
14. Ibid., 135.
15. Ibid., 137.
16. Ibid., 134.
17. Butti and Perlin, *A Golden Thread*, 140.
18. J. C. Williams, *Energy and the Making of Modern California*, 135.
19. Butti and Perlin, *A Golden Thread*, 141.
20. Ibid., 152.
21. Kapstein,"Transition to Solar."
22. Farber,"Solar Water Heating and Space Heating in Florida"; Lof and Tybout, "Cost of House Heating with Solar Energy."
23. Butti and Perlin, *A Golden Thread*, 155.
24. Spatari et al., "Twentieth Century Copper Stocks and Flows in North America," 45.
25. Butti and Perlin, *A Golden Thread*, 155.
26. Kapstein,"Transition to Solar," 114.
27. Renewable Energy Policy Network for the 21st Century, "Renewables Global Status Report 2009."
28. Junfeng and Runqing, "Solar Thermal in China," 25–27.

CHAPTER 12

1. Chase, "Sun Supplies Heat for New Type North Shore Suburb Home."
2. Lawrence Berkeley National Laboratory.
3. "'Solar House' Owner Starts Homes Project."
4. "Sun May Aid Heating of Postwar Homes."
5. "Sees Broad Demand for the 'Solar' Home."
6. "Purdue Experiments Show How Sun's Rays."

7. Simon, *Your Solar House.*

8. Ibid., 97.

9. "Solar Houses Win Approval Across Nation."

10. President's Materials Policy Commission, *Resources for Freedom,* 220.

11. Rome,"Give Earth a Chance," 51.

12. Green's Ready-Built Solar Homes.

13. Hutchinson, "The Solar House."

14. Denzer, "The Solar House in 1947," 300.

15. Rome,"Give Earth a Chance," 120.

16. Keats, *The Crack in the Picture Window,* xiv.

17. Rome,"Give Earth a Chance."

18. Usher, *A History of Mechanical Inventions,* 398.

19. Calculated from Hirsh,"PURPA," 4.

20. Hirsh,"PURPA," 5–7.

21. Hyman, *America's Electric Utilities.*

22. Hirsh,"PURPA," 9.

23. Ibid.

24. Ibid., 34.

25. Hyman, *America's Electric Utilities,* 18–19.

26. Nye, *Electrifying America,* 131.

27. Hirsh,"PURPA," 26–36.

28. Quoted in Isenstadt, "Visions of Plenty, " 313.

29. Cowan,"The 'Industrial Revolution,'" 21

30. Hirsh,"PURPA,".

31. Day, "Capital-Labor Substitution in the Home."

32. Hirsh,"PURPA," 52.

33. YouTube, "Reddy Kilowatt Commercial."

34. Hirsh,"PURPA," 52.

35. Rome,"Give Earth a Chance," 66.

36. Ibid., 59.

37. Ibid., 45–87.

38. Dwyer, Telephone interview.

39. Rome,"Give Earth a Chance," 65

40. Hirsh,"PURPA."

41. Lovins, *Soft Energy Paths,* 25.

42. Hirsh,"PURPA," 34–35, 50.

43. Rome,"Give Earth a Chance," 77.

44. Ibid., 75.

45. Dwyer, Telephone interview.
46. Hirsh,"PURPA."
47. U.S. Congress, Office for Technology Assessment, *Energy Efficiency*, 129.

CHAPTER 13

1. "The President's Solar Advocate."
2. *SERI: The First Year.*
3. "Untitled."
4. "Carter Rejects New Spending on Energy."
5. "Sun Day: 'When EXXON Gets Control.'"
6. "Untitled."
7. National Research Council, Solar Energy Research Committee. *Establishment of a Solar Energy Research Institute.*
8. Beattie, *History and Overview of Solar Heat Technologies*, 29.
9. Dooley, *U.S. Federal Investments in Energy R&D*, 10.
10. Ibid., 11.
11. Bezdek and Wendling, "A Half Century of US Federal Government Energy Incentives," 43.
12. Beattie, *History and Overview of Solar Heat Technologies*, 78.
13. "The President's Solar Advocate"
14. "Solar Activist, Denis Hayes, Heads SERI."
15. Jacobs, "Denis Hayes Is Named Director of Solar Energy Institute," 105.
16. "Sun Day Organizer Is New Solar Director."
17. Parisi, "New Path For Solar Institute."
18. "Sun Day Organizer Is New Solar Director."
19. Laird, *Solar Energy, Technology Policy, and Institutional Values*, 151.
20. "Solar Activist, Denis Hayes, Heads SERI."
21. See, for example, Atomic Energy Commissioner Edward Teller's advocation of peaceful uses of nuclear energy.
22. Berman and O'Connor, *Who Owns the Sun?*
23. Hayes, Telephone interview.
24. Ibid.
25. Ibid.
26. Young,"The Paper Mill and the City of Camas."
27. "C.W.P. Co. to Appeal $5,000 Hinz Judgment."
28. Moszeter, Interviewed by Kathy Tucker.
29. Kingsberry, Interviewed by Kathy Tucker.
30. Hayes, Telephone interview.
31. Ibid.

32. Laird, *Solar Energy, Technology Policy, and Institutional Values*, 47.
33. Hayes, Telephone interview.
34. Solar Energy Research Institute. *Institutional Plan*.
35. Hayes, Telephone interview.
36. Nemet,"Demand-Pull Energy Technology Policies."
37. Ogden, Podesta, and Deutch, "A New Strategy to Spur Energy Innovation."
38. Hayes, Telephone interview.
39. Rosenfeld, "The Art of Energy Efficiency," 51.
40. Solar Energy Research Institute. *A New Prosperity*.
41. J. Kirsch, "Book Review."
42. Bryan and McClaughry, *The Vermont Papers*, 142.
43. "Edwards Won't Confirm Energy Post."
44. Moore,"The Role of Congress," 107–111.
45. Beattie, *History and Overview of Solar Heat Technologies*, 140.
46. Hayes, Telephone interview. Also recounted in A. Allen, "Prodigal Sun."
47. Hayes, Telephone interview.
48. Rosenfeld, "The Art of Energy Efficiency," 51.
49. U.S. House of Representatives. "Economic Growth through Energy Efficiency."
50. Hayes,"Washington Decrees a Solar Eclipse."
51. Laird, *Solar Energy, Technology Policy, and Institutional Values*, 10.
52. *SERI: The First Year*, 6.

CHAPTER 14

1. Solar Energy Industries Association. "Utility-Scale Solar Projects in the United States."
2. Technically, it's a fluid known as Therminol VP-1, which is often used in making nylon and chemicals.
3. Kubiszewski,"Odeillo Font-Romeau, France."
4. Kearney, "Parabolic Trough Collector Overview."
5. "Parabolic Trough Solar Field Technology."
6. San Diego Renewable Energy Study Group. "Potential for Renewable Energy in the San Diego Region."
7. Price,"Testimony to the Public Utilities Commission of Colorado."
8. Kearney, "Parabolic Trough Collector Overview"; K. Johnson,"Solar Power" BrightSource Energy, "About Us."
9. Romm, "The Technology That Will Save Humanity."
10. Goldman, Personal interview.

11. Genesis 28:17 (New International Version).
12. Lotker, Telephone Interview.
13. Bakst,"Journey to the Secret City of Luz."
14. Reichman and Rosner,"The Bone Called Luz."
15. Reichman, "Re: Luz."
16. Goldman, *A Working Paper on Project Luz*.
17. Ibid., vi.
18. Ibid., 160.
19. Ibid., 164.
20. Ibid, 173.
21. Ibid.
22. Ibid., 154.
23. Bowes, "The Experiment That Did Not Fail."
24. For detailed descriptions of *kibbutz* life, see Mehta, "Kibbutz Life in Israel During the Late Sixties and Early- to Mid-Seventies."
25. Dar, "Communality, Rationalization and Distributive Justice," 92.
26. Goldman, *A Working Paper on Project Luz*, 154.
27. Ibid., 164.
28. Ibid., 180.
29. Goldman, Personal interview.
30. Aviel, "Effect of the World Food and Fuel Crisis on Israeli Policy-Making."
31. Berger, *Charging Ahead*, 23–48.
32. Kryza,*The Power of Light*, 3.
33. Daniels, *Direct Use of the Sun's Energy*.
34. For example, Daniels, "Selective Radiation Coatings."
35. Hallacy, *The Coming Age of Solar Energy*, 189–207; Winston and Hinterberger, "Principles of Cylindrical Concentrators for Solar Energy."
36. Stromberg, "A Status Report on the Sandia Laboratories Solar Total Energy Program," 359.
37. L. Murphy, "The Use of Solar Thermal Energy to Generate Electricity."
38. Reece, *The Sun Betrayed*, 31–32.
39. L. Murphy, "The Use of Solar Thermal Energy to Generate Electricity," 9.
40. "Tax Law Generating Big Sun-Power Sales."
41. Goldman, Personal interview.
42. Hirsh,"PURPA."
43. U.S. Congressional Senate. *National Energy Policy Act of 1989*.

44. Goldman, Personal interview.
45. Cudahy,"PURPA."
46. Goldman, Personal interview.
47. Lotker,"Barriers to Commercialization of Large-Scale Solar Electricity."
48. Ibid., 2–7.
49. Southern California Edison Company, *1987 Annual Report*, 15.
50. Lotker,"Barriers to Commercialization of Large-Scale Solar Electricity."
51. Goldman, Personal interview.
52. Lotker,"Barriers to Commercialization of Large-Scale Solar Electricity," 10.
53. Wood,"State-of-the-Art Solar Collectors For Clean Energy," 12.
54. Lotker,"Barriers to Commercialization of Large-Scale Solar Electricity," 10.
55. Goldman, Personal interview.
56. Lotker, Telephone interview.
57. Goldman, *A Working Paper on Project Luz*, 174.
58. Moore,"Solar Energy and the Reagan Administration."
59. See United States Congressional hearings in bibliography.
60. U.S. Congressional Senate, *National Energy Policy Act of 1989*, 436.
61. Yi,"Tax Boost Predicted in Face of State's Largest Deficit Since '91."
62. Richardson, "Energy Firm Pushing Hard for Property Tax Exemption."
63. Richardson, "Tax Break Quickly Ok'd for Solar Energy Plant."
64. Paparian, "Solar Power."
65. Walters, E-mail Message to author.
66. Becker, 42.
67. Gray, "Tax Break For Solar Company Gains."
68. Richardson, "Tax Break Quickly Ok'd for Solar Energy Plant."
69. Weiss, "Everyone Loves Solar Energy, But. . . ."
70. Richardson, "Tax Break Quickly Ok'd for Solar Energy Plant."
71. "Luz Casts A Long Shadow."
72. Ibid.
73. Goldman, Personal interview.
74. Parrish,"How Sun Failed to Shine on Solar Firm's Dreams."
75. U.S. Congressional House Subcommittee on Energy and Power, *Permanent Extension of Certain Expiring Tax Provisions*, 648.
76. Cable,"Solar Trough Generation."
77. Edison International, "Southern California Edison and BrightSource Energy Sign World's Largest Solar Deal."

CHAPTER 15

1. Haukos, "Fish and Wildlife Leaflet."
2. Johansen, Personal interview.
3. Ibid.
4. Sheehan, Dunahay, Benemann, and Roessler, *A Look Back at the U.S. Department of Energy's Aquatic Species Program.*
5. Knoshaug. *Current Status of the Department of Energy's Aquatic Species Program.*
6. Johansen, Personal interview.
7. Dooley, *U.S. Federal Investments in Energy R&D.*
8. Obama, "Remarks by the President at the National Academy of Sciences Annual Meeting."
9. Author's personal reporting at Google event in San Francisco.
10. US Department of Agriculture, "World Agriculture Supply and Demand Estimates Report for March 2010."
11. Cleveland, "Net Energy from the Extraction of Oil and Gas," figure 6.
12. Johansen, Personal interview.
13. Burlew, *Algal Culture,* 600.
14. Ibid., 195.
15. Ibid., 310.
16. Thacker and Babcock, "The Mass Culture of Algae," 50.
17. Sheehan, Dunahay, Benemann, and Roessler, *A Look Back at the U.S. Department of Energy's Aquatic Species Program,* i.
18. Ibid., 193.
19. Ibid., 1.
20. Darzins, E-mail to the author.
21. Ibid.
22. Ibid.
23. Deffeyes, *Hubbert's Peak,* x.
24. Kanellos, "Algae Startup #57."
25. Mouawad, "Exxon to Invest Millions to Make Fuel from Algae."
26. UTEX, The Culture Collection of Algae at the University of Texas, Austin.
27. Darzins, E-mail to the author.
28. Benemann, Guest post at R-Squared.
29. D. A. Walker, "Biofuels, Facts, Fantasy, and Feasibility."
30. Kanellos, "With Greenfuel's Demise, What Happens to Algae?"
31. Benemann, E-mail to the author.
32. Sheehan, Dunahay, Benemann, and Roessler, *A Look Back at the U.S. Department of Energy's Aquatic Species Program,* 20.

PART IV:
LESSONS FROM THE GREAT ENERGY RETHINK

CHAPTER 16

1. EIA, "Petroleum Overview, 1949–2009."
2. Buell, "Nationalist Postnationalism."
3. Rome, *The Bulldozer in the Countryside*, 120.
4. Ibid., 13, 8.
5. Bocking, *Ecologists and Environmental Politics*, 180–182.
6. Focht,"An Ecosystem Is a Partnership in Nature."
7. Archived episodes available at Thirteen,
    www.thirteen.org/ourvanishingwilderness/.
8. Schoenfeld, Meier, and Griffin, "Constructing a Social Problem," 42.
9. Perelman,"Speculations on the Transition to Sustainable Energy."
10. Rosenfeld,"The Art of Energy Efficiency."
11. Odum, *Environment, Power, and Society*, 221.

CHAPTER 17

1. V. Cohn, *1999: Our Hopeful Future*, 35.
2. There's even an award for the Council of Advancement of Science
    Writing. See http://casw.org/casw/fellowships-and-awards.
3. Schurr, *Energy, Economic Growth and the Environment*.
4. Hirsh, *Technology and Transformation*.
5. Sporn in Schurr, *Energy, Economic Growth and The Environment*, 70.
6. Ibid.
7. Hirsh, *Technology and Transformation*, 56.
8. Quoted in Lovins, *Soft Energy Paths*, 4.
9. Energy Information Administration, Energy Annual Report, figure 1.
10. R. F. Hirsh, *Technology and Transformation*, 148–155.
11. Lovins, *Soft Energy Paths*, 27.
12. Socolow,"Reflections on the 1974 APS Energy Study," 60–68.
    Rosenfeld,"The Art of Energy Efficiency," 36.
13. Rosenfeld, "Dr. Rosenfeld Video."
14. Crow and Hager. "Political Versus Technical Risk Deduction."
15. Socolow,"Reflections on the 1974 APS Energy Study," 61.
16. Ibid.
17. Rosenfeld, "A Brave New Source."
18. Rosenfeld, "The Art of Energy Efficiency."
19. "Whatever Happened to Leadership and Where Are the Leaders?"

20. R. D. Putnam, "The Strange Disappearance of Civic America."

21. Hong, Chou, and Bong, "Building Simulation," 351.

22. Rosenfeld, "The Art of Energy Efficiency."

23. Rosenfeld, "A Brave New Source."

24. Rosenfeld, "The Art of Energy Efficiency," 45.

25. Rosenfeld, with Poskanzer, "A Graph Is Worth a Thousand Gigawatt-Hours," 68.

26. Rosenfeld, "The Art of Energy Efficiency," 49.

27. Lifsher, "You Can Thank Arthur Rosenfeld for Energy Savings."

28. Sudarshan, "Deconstructing the Rosenfeld Curve," 1.

29. Ibid.

30. Rosenfeld, "The Art of Energy Efficiency," 78.

31. Kammen, December 2009 Google San Francisco event.

32. Sorrell, Dimitropoulos, and Sommerville, "Empirical Estimates of the Direct Rebound Effect," 1356.

33. Herring, for example, in the Encyclopedia of Earth. Available online at http://www.eoearth.org/article/Rebound_effect.

34. Moezzi and Diamond, "Is Efficiency Enough?" 28–38.

35. Sorrell, "Jevons' Paradox Revisited."

36. Ibid.

37. Adler, Personal communication.

38. Wei. "A General Equilibrium View of Global Rebound Effects."

39. Rutledge, "Hubbert's Peak, the Coal Question, and Climate Change."

40. Smil, *Energy at the Crossroads*, 97–105.

41. Kintisch, *Hack the Planet*.

42. V. Cohn, *1999: Our Hopeful Future*, 112.

CHAPTER 18

1. Borasi and Zardini, *Out of Gas*, 178, 179.

2. Ibid.

3. Skurka and Naar, *Design for a Limited Planet*, 92.

4. DeKorne, *The Survival Greenhouse*.

5. Ibid.

6. Skurka and Naar, *Design for a Limited Planet*, 92.

7. Ibid.

8. Sylla, "The U.S. Banking System."

9. Paulding, *The New Mirror for Travelers*, 5.

10. Stoehr, *Nay-Saying in Concord*, 16.

11. Rose, *Transcendentalism as a Social Movement*.

12. Thoreau, "Paradise (to be) Regained," 460.

13. Thoreau, *Walden*, 6.

14. Borasi and Zardini, *Out of Gas*, 89.

15. Skurka and Naar. 57.

16. Ibid., 53.

17. Ibid., 158.

18. Ibid., 130.

19. Ibid., 75–76.

20. Borasi and Zardini, *Out of Gas*.

21. DeKorne, *The Survival Greenhouse*, back cover.

22. Naar, E-mail to the author.

23. Reich, *The Greening of America*, 350.

24. Ibid., 363.

25. Thoreau, *Walden*, 10.

26. Schiff, "Neo-Transcendentalism in the New Left Counter-Culture."

27. Ibid.

28. Rome, "Give Earth a Chance."

29. Carson, *Silent Spring*.

30. Commoner, *The Poverty of Power*, 5.

31. DeKorne, *The Cracking Tower*, xx.

32. DeKorne, *Survival Greenhouse*, 9.

33. Kirk, *Counterculture Green*, 97.

34. Schell, *The Town That Fought to Save Itself*.

35. Gutowski et al., "Environmental Life Style Analysis (ELSA)."

36. DeKorne, *Survival Greenhouse*, 8.

37. DeKorne, *Cracking Tower, 39*.

38. Mattson, *What the Heck Are You Up To, Mr. President?*, 36.

CHAPTER 19

1. Brand, *The Next Whole Earth Catalog*, 133.

2. Baer, *Sunspots*, 25.

3. Ibid.

4. Kirk, *Counterculture Green*, 54.

5. Baldwin and Brand, *Soft-tech*, 4–5.

6. Ibid., 5.

7. Schumacher, *Small Is Beautiful*, 29.

8. Ibid., 13.

9. Ellul, *The Technological Society*, 17.

10. Mumford, *The Pentagon of Power*, 395–396.

11. Meadows, "Let's Have a Little More Feedback"
12. Odell, *Environmental Awakening*, 191.
13. Brand, *Whole Earth Discipline*.
14. Righter, *Wind Energy in America*, 93–100.
15. Ibid., 101.
16. Baldwin and Brand, *Soft-tech*, 1.
17. Bookchin, *Post-Scarcity Anarchism*, 68.
18. Baer, *Sunspots*, 5–7.
19. Laird, "Constructing the Future," 44.
20. Kirk, *Counterculture Green*, 17.
21. Baer, *Sunpsots*, 53.
22. Winner, *The Whale and the Reactor*, 80.
23. Harper, "Transfiguration among the Windmills," 14.
24. Ibid., 13.
25. Winner, *The Whale and the Reactor*, 82.
26. Lee and Donnelly, *Solar Failure*, iv.
27. Baldwin.
28. Skurka and Naar, *Design for a Limited Planet*, 192–197.
29. Righter, *Wind Energy in America*, 188.
30. Weil, "The Wind Farmers of East 11th Street."
31. Baldwin and Brand, *Soft-tech*, 1.
32. Bethell, Tom. *The Electric Windmill*, 105–107.
33. Kirk, *Counterculture Green*, 31.

CHAPTER 20

1. Perlin, *From Space to Earth*, 52–55.
2. Nemet, "Beyond the Learning Curve."
3. Lovins, *Soft Energy Paths*, 143.
4. Brand, *Next Whole Earth Catalog*, 195.
5. Reece, *The Sun Betrayed*, 77.
6. Ibid., 167.
7. "Harold R. Hay: Solar Pioneer."
8. Kazmerski, " Solar Photovoltaics R&D at the Tipping Point," 106.
9. Quoted in Hallacy, *The Coming Age of Solar Energy*.
10. "Solar Energy Method Found."
11. Hallacy, *The Coming Age of Solar Energy*, 202.
12. Ibid., 193–195.
13. Meinel, Telephone interview.

14. Hammond and Metz, "Solar Energy Research."

15. Ibid., 241.

16. Turner, *From Counterculture to Cyberculture.*

PART V:
INNOVATIONS AND THE FUTURE

CHAPTER 21

1. G. Macdonald, "Earth Day and Coal."

2. F. U. Adams,"Report of the Bureau of Smoke Inspection," 191.

3. Hardin, "The Tragedy of the Commons," 1245.

4. Cowling,"Acid Precipitation in Historical Perspective."

5. EIA, "Elecric Power Annual, 2009," table 3.9.

6. For example, see Dooge, *An Agenda of Science for Environment and Development*, 144.

7. Weart, *The Discovery of Global Warming.*

8. Saunders, *Compressed Air Information.*

9. G. Macdonald, "Earth Day and Coal."

10. Nemet, "Beyond the Learning Curve."

11. White, *The Organic Machine*, 6.

12. Sinclair, *Philadelphia's Philosopher Mechanics*, 144–146.

13. Lewis, *The Romance of Water-power.*

14. White, *The Organic Machine*, 7.

CHAPTER 22

1. "Harnessing the Wind."

2. P. C. Putnam, *Power from the Wind.*

3. Ibid., 13.

4. Bush, *Pieces of the Action*, 105.

5. Righter, *Wind Energy in America.*

6. P. C. Putnam, *Power from the Wind.*

7. "G. H. Putnam Left Estate of $675,540."

8. "P. C. Putnam Bankrupt."

9. P. C. Putnam, "A Reconnaissance Among Some Volcanoes."

10. P. C. Putnam, *Power from the Wind*, 1-2.

11. Ibid., 3.

12. Ibid., 5.

13. Righter, *Wind Energy in America*, 34.

14. P. C. Putnam, *Power from the Wind*, vi.

15. Righter, *Wind Energy in America*, 74.

16. Stewart, Interviewed by John L. Greenberg.

17. Hiester and Pennell, "The Meterological Aspects of Siting Large Wind Turbines."

18. P. C. Putnam, *Power from the Wind*, 36.

19. Ibid., 43.

20. Ibid., xi.

21. Twain, *Connecticut Yankee in King Arthur's Court*, 5.

22. P. C. Putnam, *Power from the Wind*, 12.

23. Ibid., 13.

24. Righter, *Wind Energy in America*, 143.

25. Savino, "A Brief Summary of the Attempts to Develop Large Wind-Electric Generating Systems in the U.S."

26. Woody,"Wind Jobs Outstrip Coal Mining."

CHAPTER 23

1. Wicker,"Johnson Reports a 'Breakthrough' in Atomic Power."

2. L. B. Johnson,"Commencement Address at Holy Cross College."

3. Ibid.

4. Ibid.

5. Ibid.

6. EIA, "Generation by Energy Source."

7. Laird.

8. Lovins, *Soft Energy Paths*.

9. "Blunt Talk to Solar Energy Supporters."

10. Weinberg, "Can the Sun Replace Uranium?"

11. Daniels,"Neutronic Reactor."

12. Brand.

13. Rosenfeld,"A Brave New Source."

14. Weinberg, "Can the Sun Replace Uranium?" 27.

15. Rashad and Hammad, "Nuclear Power and the Environment."

16. J. W. Johnson, *Insuring Against Disaster*, 64.

17. S. Cohn, *Too Cheap to Meter*, 24.

18. Ibid., 22.

19. Wicker,"Johnson Reports a 'Breakthrough' in Atomic Power."

20. "Coal Industry Urges New Look at A-Energy."

21. S. Cohn, *Too Cheap to Meter*, 32.

22. Kotulak, "Scientists Tell of Hopes for New Atom Plant."

23. Weinberg, "Can the Sun Replace Uranium?"
24. Kotulak, "Scientists Tell of Hopes for New Atom Plant."
25. S. Cohn, *Too Cheap to Meter*, 32.
26. Bishop,"Cheaper Power."
27. Toth, "U.S. A-Plant Estimates Challenged by Russian."
28. Weinberg, *The First Nuclear Era*, 135.
29. For example, see Finney, "Nuclear-Power Costs Reduced as Atomic Fuels Exceed Their Life Expectancy."
30. Simons,"Dispute Flares Over A-Power Cost Dip."
31. S. Cohn, *Too Cheap to Meter*, 105.
32. "Atoms for Peace."
33. Smil, *Energy at the Crossroads*, 141.
34. Seaborg, *Peaceful Uses of Nuclear* Energy, 3.
35. Ibid., 4.
36. Weinberg, *The First Nuclear Era*.
37. S. Cohn, *Too Cheap to Meter*.
38. Ibid., 333.
39. Wellock, *Critical Masses*.
40. Gilinsky, "Nuclear Safety Regulation Lessons from US Experience," 707.
41. Weinberg, *The First Nuclear Era*, 153–175.
42. S. Cohn, *Too Cheap to Meter*.
43. Smil, *Energy at the Crossroads*.
44. Hyman, *America's Electric Utilities*, 110.
45. Harris, *Climate Change and American Foreign Policy*.
46. Obama,"Remarks by the President Challenging Americans to Lead the Global Economy in Clean Energy."
47. McVeigh et al., "Winner, Loser, or Innocent Victim?"
48. Weinberg, *The First Nuclear Era*, 275.

CHAPTER 24

1. Wald, "A New Era for Windmill Power."
2. Asmus, *Reaping the Wind*.
3. "Breezing into the Future."
4. "1993 Discover Awards."
5. American Wind Energy Association, "FAQ."
6. Asmus, *Reaping the Wind*, 158.
7. Ibid., 160.
8. Ibid.
9. Wald, "A New Era for Windmill Power."

10. "The Green Factor."
11. Ibid.
12. Kamp, Ruud, and Andriesse, "Notions on Learning Applied to Wind Turbine Development"; Ibenholt, "Explaining Learning Curves for Wind Power."
13. Hirsh, "PURPA," 168–169.
14. Asmus, *Reaping the Wind.*
15. Pursell," The Rise and Fall of the Appropriate Technology Movement," 632, 633.
16. Asmus, *Reaping the Wind,* 116.
17. Ibid., 76.
18. Nemet, "Demand-Pull Energy Technology Policies."
19. Asmus, *Reaping the Wind,* 91.
20. D. R. Smith, "The Wind Farms of Altamont Pass," 146.
21. Asmus, *Reaping the Wind,* 122.
22. "Community Choice Laws and the Inevitable Solar-Hydrogen Econ."
23. Righter, *Wind Energy in America.*
24. Asmus, *Reaping the Wind,* 72.
25. D. R. Smith, "Wind Farms of Altamont Pass," 145.
26. Nemet, "Demand-Pull Energy Technology Policies."
27. Asmus, *Reaping the Wind,* 87–88.
28. Ibid., 124.
29. Ibid.
30. Brower, Personal interview.
31. Felker, Telephone interview.
32. Nierenberg,"Free-Flow Variability on the Jess and Souza Ranches, Altamont Pass," 50.
33. Hiester and Pennell, "The Meterological Aspects of Siting Large Wind Turbines"; "Wind Energy Resource Atlas."
34. Asmus, *Reaping the Wind,* 75.
35. D. R. Smith, "Wind Farms of Altamont Pass," 161.
36. Lesser and Su, "Design of an Economically Efficient Feed-in Tariff Structure."
37. Rowlands, "Envisaging Feed-In Tariffs for Solar Photovoltaic Electricity."
38. European Parliamentary Temporary Committee on the Echelon Interception System, "Report on the Existence of a Global System for the Interception of Private and Commercial Communications."
39. Asmus, *Reaping the Wind,* 147.
40. Stover,"The Forecast for Wind Power."

41. Asmus, *Reaping the Wind*, 124.

42. Kamp, Ruud, and Andriesse, "Notions on Learning Applied to Wind Turbine Development," 1632.

43. Karnoe, "Technological Innovation and Industrial Organization in the Danish Wind Industry," 284–291.

44. D. R. Smith, "The Wind Farms of Altamont Pass," 151.

45. Gipe,"Wind Energy Comes of Age," 85.

46. Stover,"The Forecast for Wind Power," 69.

47. Asmus, *Reaping the Wind*, 189.

48. *Power-Electronic, Variable-Speed Wind Turbine Development.*

49. Berger, *Charging Ahead*, 165.

50. "Share Price Volatile After Sell Direction Shakes Investors."

51. Ibid.

52. Ibid.

53. Ibid.

54. "Pertinent Facts of the Kenetech Fiasco."

55. Nemet, "Demand-Pull Energy Technology Policies."

56. Mortensen, "International Experiences of Wind Energy."

57. Gipe, "Design as if People Matter."

58. Thayer, "Altamont," 379.

59. Hoen et al., "The Impact of Wind Power Projects on Residential Property Values."

60. Dehlsen, "Wind Power Pioneer Interview: Jim Dehlsen, Clipper Windpower."

61. Garud and Karnoe,"Bricolage Versus Breakthrough."

62. Felker, Telephone interview.

63. Karnoe, "Technological Innovation and Industrial Organization in the Danish Wind Industry."

64. *The Globe and Mail*, August 31, 1992.

65. "Breezing into the Future," 48.

66. American Wind Energy Association. "Production Tax Credit."

67. Vietor, *Energy Policy in America Since 1945*, 228.

68. Green, "The Elusive Green Economy."

CHAPTER 25

1. EPRI, *Energy Storage in a Restructured Electric Utility Industry.*

2. Moutoux, "Wind Integrated Compressed Air Storage in Colorado."

3. Marcus, Telephone interview..

4. General Compression, "General Compression Closes on $17 Million."

5. Pickard, Shen, and Hansing, "Parking the Power."
6. Zahner, *The Transmission of Power by Compressed Air*, 13.
7. EIA, "Weekly Natural Gas Storage Report."
8. Mitchell,"Power Plant Uses Compressed Air to Make Energy."
9. Nakhamkin, Personal interview.
10. Ibid.
11. EPRI, *Energy Storage in a Restructured Electric Utility Industry*.
12. Madrigal, "DOE Report Says More Wind Than Coal Planned for US Grid."
13. Greenblatt et al., "Baseload Wind Energy."
14. Cavallo,"Controllable and Affordable Utility-Scale Electricity Generation," 123.
15. Kraemer, Susan. "California Gets Smart-Grid Funds to Bottle Wind."
16. Garud and Karnoe, "Bricolage Versus Breakthrough."

CHAPTER 26

1. Stoll, *The Great Delusion*, 99.
2. Etzler, *The Paradise Within the Reach of All Men*.
3. Stoll, *The Great Delusion*.
4. Butti and Perlin, *A Golden Thread*, 34-37.
5. Ibid., 34.
6. Etzler, *The Paradise Within the Reach of All Men*, 34.
7. Mollan,"Stopping the Sun," 75.
8. "Archimedes Proved to Be a Dazzling Warrior."
9. Hughes, *Human-Built World*.
10. Google, "Google's Goal."
11. LaMonica, "Q&A."
12. Tuomi, "The Lives and the Death of Moore's Law."
13. Hughes, *Human-Built World*, 101.
14. Ibid., 101.
15. Hirsh, *Technology and Transformation in the American Electric Utility Industry*, 67.
16. Gross, Telephone interview.
17. Fishman, Telephone interview.
18. Areva, "Areva Acquires the U.S. Solar Company Ausra."
19. Sass, *The Substance of Civilization*, 4.
20. Hirsh, *Technology and Transformation in the American Electric Utility Industry*, 92-93.
21. Goetzberger et al., "Solar Cells."

22. Kazmerski, "Solar Photovoltaics R&D at the Tipping Point," 105–135.
23. Ceder, *The Materials Genome.*
24. Waldau-Jager, "Status of Thin Film Solar Cells in Research, Production, and Market.".
25. Delgass et al., "Report on the Workshop on a Drug Discovery Approach," 5.
26. Ceder, Telephone interview.
27. Ibid.
28. Ibid.
29. Kang and Ceder, "Battery Materials for Ultrafast Charging and Discharging."
30. Kaika, "Dams as Symbols of Modernization."
31. Pimental et al., "Renewable Energy."
32. Huet, *Memoirs of the Life of Peter Daniel Huet*, 211.
33. Kristensen, "Fragments of the Cup Anemometer History."
34. Ibid., 2.
35. Pacific Northwest National Laboratory, "Meteorological Siting."
36. Brower, Personal interview.
37. Garud and Karnoe, "Bricolage Versus Breakthrough."
38. Baker, *A Field Guide to American Windmills*, 210.
39. "Catch the Wind Whitepaper."
40. Ibid., 1.
41. "Smarting from the Wind."
42. Felker, Personal interview.
43. Reich, *The Greening of America.*
44. Fetzer, Personal interview.
45. Crutzen,"The 'Anthropocene.'"

CHAPTER 27

1. Bathel, "The Biogeography of the Desert Tortoise."
2. National Renewal Energy Laboratory, "Photovotalic Visualization."
3. California Energy Commission, Evidentiary Hearing, January 11, 2010, 367.
4. Cronon,"The Trouble with Wilderness."
5. California Energy Commission, Evidentiary Hearing, January 11, 2010, 3.
6. Ibid., 2.
7. See California Energy Commission, Docket Number 07-AFC-05, http://www.energy.ca.gov/sitingcases/ivanpah/documents/index .html.

8. Barringer, "Environmentalists in a Clash of Goals."
9. Navarro, "Green Scene."
10. P. C. Murphy, *What a Book Can Do*.
11. S. Adams, "Environmental Issues and Politics."
12. Odell, *Environmental Awakening*.
13. Carson, *Silent Spring*, 297.
14. Schumacher, *Small Is Beautiful*, 31.
15. Jasper and Poulsen, "Recruiting Strangers and Friends."
16. Skurka and Naar, *Design for a Limited Planet*, 175.
17. McGovern, "Introduction," xi.
18. Ehrlich, *Population Bomb*, 61.
19. Ibid., 77.
20. Nordhaus and Shellenberger, *Break Through*.
21. Frank, "Science, Nature, and the Globalization of the Environment."
22. Jasper and Poulsen, "Recruiting Strangers and Friends."
23. Rome, Telephone interview.
24. Kirk, *Counterculture Green*, 16.
25. Galbraith, *The Affluent Society*, 187.
26. Rome, "Give Earth a Chance."
27. California Energy Commission, Evidentiary Hearing, January 11, 2010, 209.
28. Rome, "Give Earth a Chance," para. 74.
29. Ibid.
30. California Energy Commission, Evidentiary Hearing, January 12, 2010, 14.
31. California Energy Commission, Evidentiary Hearing, January 11, 2010.
32. Jasper, *The Art of Moral Protest*, 78.
33. Ellis, "Anthropogenic Biomes in the Global Ecosystem."
34. California Energy Commission, "Staff's Opening Brief," April 1, 2010.
35. Hsu.
36. Ellis, Erle "Op-Ed: Stop Trying to Save the Planet."
37. Cronon, "The Trouble with Wilderness," 80.

BIBLIOGRAPHY

"1993 Discover Awards: Environment: Reaping the Wild Wind." *Discover.*
October 1993.

Abernathy, Lloyd M. "Progressivism, 1905–1919." In *Philadelphia: A 300
Year History,* ed. Russell Frank Weigley and Edwin Wolf, 524–565.
New York: W. W. Norton, 1982.

Adams, Edward Dean. *Niagara Power: History of the Niagara Falls Power
Company (1886–1918).* Niagara Falls, NY: Niagara Falls Power
Company, 1927.

Adams, Sean. "Environmental Issues and Politics." *The Princeton
Encyclopedia of American Political History,* ed. Michael Kazin et al.,
297–300. Princeton, NJ: Princeton University Press, 2010.

Adler, Dan. Personal communication. March 2010.

Allen, Arthur. "Prodigal Sun." *Mother Jones.* March/April 2000.
http://motherjones.com/politics/2000/03/prodigal-sun.

Allen, Frederick Lewis. *The Big Change: America Transforms Itself, 1900–
1950.* New York: Harper, 1952.

American Wind Energy Association. "Production Tax Credit."
http://www.awea.org/ei_policy_ptc.cfm.

———. "FAQ." www.awea.org/faq/wwt_basics.html.

Appleton, Joseph. "Recent Developments in the Applications of Storage
Batteries." *Proceedings of the Engineers' Club of Philadelphia* 15 (March
1998), 178–200.

"Archimedes Proved to Be a Dazzling Warrior." *New York Times* News
Service. *The Milwaukee Journal.* December 24, 1973.

Areva. "Areva Acquires the U.S. Solar Company Ausra." Press release.
February 9, 2010. www.areva.com/EN/news-8199/areva-to-acquire-
the-u-s-solar-company-ausra.html.

Aronson, Sidney H. "The Sociology of the Bicycle." *Social Forces* 30, no. 3
(March 1952): 305–312.

"A Smokeless City." *Los Angeles Herald.* August 18, 1907, 7, image 7.

Asmus, Peter. *Reaping the Wind: How Mechanical Wizards, Visionaries, and Profiteers Helped Shape Our Energy Future*. Washington, DC: Island Press, 2001.

"Atoms for Peace." Eisenhower Archives online. http://www.eisenhower .archives.gov/All_About_Ike/Speeches/Atoms_for_Peace.pdf.

Aughey, Samuel. *Sketches of the Physical Geography and Geology of Nebraska*. Omaha, NE: Daily Republican Book and Job Office, 1880.

Aviel, JoAnn Fagot. "Effect of the World Food and Fuel Crisis on Israeli Policy-Making." *Political Research Quarterly* 31, no. 3 (1978): 317–333.

"A Wife's Appeal." *Daily Alta* (California). February 25, 1888. (See http://cdnc.ucr.edu/cdnc/cgi-bin/cdnc?a=d&cl=search&d= DAC18880225.2.57&srpos=10&e=———en-Logical-20—1-by TI—-boggs-all—-1888.)

Ayers, Andrew. *The Architecture of Paris: An Architectural Guide*. Stuttgart, Germany: Edition Axel Menges, 2004.

Ayres, Andrew U., and Benjamin Warr. "Dematerialization vs. Growth. Is It Possible to Have Our Cake and Eat It?" Working Paper. 2004.

Baer, Steve. *Sunspots: Collected Facts and Solar Fiction*. Albuquerque, NM: Zomeworks Corporation, 1977.

Bailey, William J. "Solar Heater." U.S. Patent 966070. August 2, 1910.

Baker, T. Lindsay. *A Field Guide to American Windmills*. Norman: University of Oklahoma Press, 1985.

Bakst, Joel David. "Journey to the Secret City of Luz." City of Luz. 2006. http://www.chazonhatorah.org/journey-to-the-secret-city-of-luz-metatron-metatron-jacob-s-ladder-to-the-face-of-god.htm.

Baldwin, J. and Brand, Stewart. *Soft-tech*. New York: Penguin, 1978.

Barbour, Erwin. *Report of the State Geologist*. Lincoln, NE: Jacob North & Co, 1903.

Barringer, Felicity. "Environmentalists in a Clash of Goals." *New York Times*. March 23, 2009. www.nytimes.com/2009/03/24/science/earth/24ecowars.html.

Bathel, Kerrie. "The Biogeography of the Desert Tortoise." San Francisco State University Department of Geography. Fall 2000. http://bss.sfsu.edu/geog/bholzman/courses/Fall00Projects/ tortoise.htm.

Beattie, Donald, ed. *History and Overview of Solar Heat Technologies*. Cambridge, MA: MIT Press, 1997.

Bellamy, Edward. *Looking Backward: 2000–1887*. New York: Modern Library, 1917.

Benemann, John. E-mail to the author. December 11, 2009.

———. Guest post at R-Squared. Consumer Energy Report. www.consumerenergyreport.com/2007/05/14/algal-biodiesel-fact-or-fiction/.

Bennett, John E. "Wave-Motors." *McBride's Magazine* 71 (January–June 1903): 251–255. Philadelphia, PA: J. P. Lippincott Co.

Berger, John J. *Charging Ahead: The Business of Renewable Energy and What It Means for America*. Berkeley: University of California Press, 1998.

Berman, Daniel M. and John T. O'Connor. *Who Owns the Sun?: People, Politics, and the Struggle for a Solar Economy*. White River Junction, VT: Chelsea Green Publishing Company, 1996.

Bethell, Tom. *The Electric Windmill: An Inadvertent Autobiography*. Washington, DC: Regnery Gateway, 1988.

Bezdek, Roger H., and Robert M. Wendling. "A Half Century of US Federal Government Energy Incentives: Value, Distribution, and Policy Implications." *International Journal of Global Energy Issues* 27, no. 1 (2007): 42–60.

"Bill Gates on Energy: Innovating to Zero!" TED.com, February 2010. http://www.ted.com/talks/lang/eng/bill_gates.html.

Bishop, Jerry E. "Cheaper Power." *Wall Street Journal*. July 20, 1964.

Black, Brian, *Petrolia: The Landscape of America's First Oil Boom*. Baltimore, MD: Johns Hopkins Press, 2000.

"Blunt Talk to Solar Energy Supporters." *Science News*. February 2, 1974.

Bocking, Stephen. *Ecologists and Environmental Politics: A History of Contemporary Ecology*. New Haven, CT: Yale University Press, 1997.

Bookchin, Murray. *Post-Scarcity Anarchism*. Montréal: Black Rose Books, 1986.

Borasi, Giovann, and Mirko Zardini. *Out of Gas*. Museum catalog. Montreal: Canadian Centre for Architecture, 2007.

Bowes, Alison M. "The Experiment That Did Not Fail: Image and Reality in the Israeli Kibbutz." *International Journal of Middle East Studies* 22 (1990): 85–104.

Boyer, Paul S., Clifford Clark, Karen Halttuenen, Joseph F. Kett, and Neal Salisbury. *The Enduring Vision: A History of the American People*. Eds. Boston: Wadsworth, 2010.

Brand, Stewart, ed. *The Next Whole Earth Catalog*. New York: Point/Random House, 1980.

Brand, Stewart. *Whole Earth Discipline: An Ecopragmatist Manifesto*. New York: Viking, 2009.

"Breezing into the Future." *Time*. January 13, 1992.

Brierley, Benjamin. *Ab-o'th'-Yate in Yankeeland*. Manchester: A. Heywood & Son, 1885. http://library.uml.edu/clh/All/bbri.htm.

BrightSource Energy. "About Us." www.brightsourceenergy.com/about_us/.

Brimblecombe, Peter. "London Air Pollution, 1500–1900." *Atmospheric Environment* 11, no. 12 (1977): 1157–1162.

Brower, Michael. Personal interview. February 18, 2010.

Bryan, Frank, and John McClaughry. *The Vermont Papers: Recreating Democracy on a Human Scale*. White River Junction, VT: Chelsea Green Publishing, 1989.

Buell, Frederick. "Nationalist Postnationalism: Globalist Discourse in Contemporary American Culture." *American Quarterly* 50, no. 3 (September 1998): 551–553.

Burlew, John S., ed. *Algal Culture: From Laboratory to Pilot Plant*. Washington, DC: Carnegie Institution, 1953.

Bush, Vannevar. *Pieces of the Action*. New York: William Morrow & Co., 1970.

Butti, Ken and Perlin, John. *A Golden Thread*. Palo Alto, CA: Cheshire Books, 1980.

Cable, Robert. "Solar Trough Generation: The California Experience." A presentation at the American Solar Energy Society 2001. *http://www.nrel.gov/csp/troughnet/pdfs/cable_frier_calexpr.pdf*.

California Energy Commission. Evidentiary Hearing. January 11, 2010. Docket Number 07-AFC-05.

———. Evidentiary Hearing. January 12, 2010. Docket Number 07-AFC-05.

———. "Staff's Opening Brief." April 1, 2010. Docket No. 07-AFC-5.

Campbell-Kelly, Martin. "Not Only Microsoft: The Maturing of the Personal Computer Software Industry, 1982–1995," *Business History Review* 75, no. 1, Computers and Communications Networks (Spring 2001): 128–134.

Carson, Rachel. *Silent Spring*. Boston: Houghton Mifflin, 2002.

"Carter Rejects New Spending on Energy." *Los Angeles Times*. May 3, 1978.

Case, Mrs. L. J. B. "Notes from a Journal, No. III." *The Ladies' Repository* 14 (1846): 181–183. Boston: A. Tompkins.

"Catch the Wind Whitepaper." Catch the Wind. http://www.catchthewindinc.com/files/NPPD%20Published%20 Report-FINAL.pdf.

Cavallo, Alfred. "Controllable and Affordable Utility-Scale Electricity Generation from Intermittent Wind Resources and Compressed Air Storage (CAES)." *Energy* 32 (2007): 120–127.

Ceder, Gerbrand. *The Material Genome*. An unpublished draft plan.

———. Telephone interview. January 5, 2010.

Chase, Al. "Sun Supplies Heat for New Type North Shore Suburb Home." *Chicago Daily Tribune*. September 22, 1940.

Cleveland, Cutler J. "Net Energy from the Extraction of Oil and Gas in the United States." *Energy* 30, no. 5 (April 2005): 769–782.

"Coal Industry Urges New Look at A-Energy." Associated Press. *Washington Post*. March 5, 1964, A2.

Cohn, Steven. *Too Cheap to Meter*. Albany: State University of New York Press, 1997.

Cohn, Victor. *1999: Our Hopeful Future*. Indianapolis, IN: The Bobbs-Merrill Company, 1956.

Cole, Arthur. "Underground Social Capital." *The Business History Review* 42, no. 4 (Winter 1968): 482–492.

Commoner, Barry. *The Poverty of Power*. New York: Bantam Books, 1977.

"Community Choice Laws and the Inevitable Solar-Hydrogen Econ." YouTube. Posted January 20, 2008. http://www.youtube.com/watch?v=BcHG8jePdyQ.

Cowan, Ruth Schwartz. "'The "Industrial Revolution"' in the Home: Household Technology and Social Change in the 20th Century," *Technology and Culture* 17, no. 1 (January 1976): 1–23.

Cowling, E. B. "Acid Precipitation in Historical Perspective." *Environmental Scientific Technology* 16, no. 2 (1982): 110A–123A.

Crockett, Davy. *An Account of Col. Crockett's Tour to the North and Down East*. Philadelphia: E. L. Carey and A. Hart, 1835.

Cronon, William. "The Trouble with Wilderness: Or, Getting Back to the Wrong Nature." In *Uncommon Ground: Rethinking the Human Place in Nature*, ed. William Cronon, 69–90. New York: W. W. Norton & Co., 1995.

Crow, Michael M., and Gregory L. Hager. "Political Versus Technical Risk Deduction and the Failure of U.S. Synthetic Fuel Development Efforts." *Review of Policy Research* 5, no. 1 (August 1985): 145–152.

Crutzen, Paul. "The 'Anthropocene.'" *Earth System Science in the Anthropocene.* Berlin: Springer, 2006.

Cudahy, Richard D. "PURPA: The Intersection of Competition and Regulatory Policy." *Energy Law Journal* 16 (1995): 419–439.

Cumbler, John T. "Review of *Nature Incorporated: Industrialization and the Waters of New England.*" *Journal of Social History* 26, no. 3 (1993): 625–626.

"C.W.P. Co. to Appeal $5,000 Hinz Judgment." *The Camas Post.* December 9, 1932.

Dalton, G. J., R. Alcorn, and T. Lewis. "Case Study Feasibility Analysis of the Pelamis Wave Energy Convertor in Ireland, Portugal and North America." *Renewable Energy* 35 (2010): 443–455.

Dalzell, Robert F. *Enterprising Elite: The Boston Associates and the World They Made.* Cambridge, MA: Harvard University Press, 1987.

Daniels, Farrington. *Direct Use of the Sun's Energy.* New Haven, CT: Yale University Press, 1964.

———. "Neutronic Reactor." U.S. Patent number 3069341, 1962.

———. "Selective Radiation Coatings: Preparation and High-temperature Stability." *Solar Energy* 3, no. 3 (October 1959): 2–7.

Dar, Yechezkel. "Communality, Rationalization and Distributive Justice: Changing Evaluation of Work in the Israeli Kibbutz." *International Sociology* 17 (2002): 91–111.

Darzins, Al. E-mail to the author. December 21, 2009.

Day, Tanis. "Capital-Labor Substitution in the Home." *Technology and Culture* 33, no. 2 (April 1992): 302–327.

Deffeyes, Kenneth. *Hubbert's Peak: The Impending World Oil Shortage.* Princeton. NJ: Princeton University Press, 2009.

Dehlsen, Jim. "Wind Power Pioneer Interview: Jim Dehlsen, Clipper Windpower." U.S. Department of Energy. October 1, 2003. http://wpadev.nrel.gov/filter_detail.asp?itemid=683.

DeKorne, James B. *The Survival Greenhouse: An Eco-System Approach to Home Food Production.* Santa Fe, NM: Blue Feather Press. 1975.

———. *The Cracking Tower: A Strategy for Transcending 2012.* Berkeley, CA: North Atlantic Books, 2009.

Delgass, W. Nicholas, James M. Caruthers, Stephen D. Stamatis, Donald R. Sadoway, Luis A. Ortiz, and Gerbrand Ceder. "Report on the Workshop on a Drug Discovery Approach to Breakthroughs in Batteries." Workshop held September 8–9, 2008 at MIT. http://web.mit.edu/dsadoway/www/NSFBatteryReport%20Final.pdf.

Denzer, A. "The Solar House in 1947." In Eco-Architecture II: Harmonisation between Architecture and Nature (Transactions on Ecology and the Environment series, vol. 113), ed. G. Broadbent and C. A. Brebbia, 295–304. Southampton and Boston: WIT Press, 2008.

"Destitute of Meaning." San Francisco Chronicle. February 27, 1898.

Dickens, Charles. Oliver Twist. London: Richard Bentley, 1839.

Dooge, James. An Agenda of Science for Environment and Development into the 21st Century. Cambridge: University of Cambridge Press, 1992.

Dooley, J. J. U.S. Federal Investments in Energy R&D: 1961–2008. Richland, WA: Pacific Northwest National Laboratory, 2008.

Duffy, Terrance. "Wave-Power Air-Compressor." U.S. Patent No. 547338, 1895.

Duncan, Louis. "The Future of Electricity in Railroad Work." Journal of the Franklin Institute 141, nos. 841–846 (January–June 1896): 401–416.

Dwyer, Polly. Telephone interview. November 6, 2009.

"Earth's Boundaries?" Nature 461 (September 24, 2009): 447–448. www.nature.com/nature/journal/v461/n7263/full/461447b.html.

Eaton, Samuel John Mills. Petroleum: A History of the Oil Region of Venango County, Pennsylvania. Philadelphia, PA: J. P. Skelly & Co., 1866.

Economides, Nicholas. "Competition and Vertical Integration in the Computing Industry." In Competition, Innovation, and the Role of Antitrust in the Digital Marketplace, ed. Jeffrey A. Eisenach and Thomas M. Lenard, 209–216. New York: Kluwer Academic Publishers, 1998.

Edgerton, David. The Shock of the Old: Technology and Global History Since 1900. Oxford: Oxford University Press, 2007.

Edison International, "Southern California Edison and BrightSource Energy Sign World's Largest Solar Deal," February 11, 2009. http://www.edison.com/pressroom/pr.asp?id=7174.

"'Edison Said Only a Few Years Since' [advertisement]." Los Angeles Herald, September 05, 1907, 4, image 4.

"Editorial: Letter IX." *Southern Literary Messenger* II, no. 57 (1835–1836): 445–450. Richmond, VA: T. W. White.

Edwards, Jim and Wynette Edwards. *Batavia: From the Collection of the Batavia Historical Society (Images of America)*. Chicago: Arcadia, 2000.

"Edwards Won't Confirm Energy Post." Associated Press. December 18, 1980.

Ehrlich, Paul. *Population Bomb*. New York: Ballantine, 1968.

EIA (U.S. Energy Information Administration). "Electric Power Annual, 2009." http://www.eia.doe.gov/cneaf/electricity/epa/epa_sprdshts .html.

———. "Generation by Energy Source." http://www.eia.doe.gov/cneaf/electricity/epm/table1_1.html.

———. "Petroleum Overview, 1949–2009." http://www.eia.gov/aer/txt/ptb0501.html.

———. "State Energy Profiles." http://tonto.eia.doe.gov/state/.

———. "Weekly Natural Gas Storage Report." November 24, 2010. http://ir.eia.doe.gov/ngs/ngs.html.

Ellis, Erle. "Anthropogenic Biomes in the Global Ecosystem." 94th ESA Annual Meeting. August 2–7, 2009. Albuquerque, NM. http://eco.confex.com/eco/2009/techprogram/P15960.HTM.

———. "Op-Ed: Stop Trying to Save the Planet." *Wired*. May 6, 2009. www.wired.com/wiredscience/2009/05/ftf-ellis-1/.

Ellul, Jacques. *The Technological Society: A Penetrating Analysis of Our Technical Civilization and of the Effect of an Increasingly Standardized Culture on the Future of Man*. New York: Vintage, 1967.

Engerman, Stanley Lewis, ed. *Cambridge Economic History of the United States: The Colonial Era*. Cambridge: Cambridge University Press, 2000.

Epperson, Bruce. "Failed Colossus: Strategic Error at the Pope Manufactory Company 1878–1900." *Technology and Culture* 41, no. 2 (April 2000): 300–320.

EPRI. *Energy Storage in a Restructured Electric Utility Industry: Report on EPRI Think Tanks I and II*. TR-108894. Palo Alto, CA: EPRI, 1997.

Etzler, John A. *The Paradise Within the Reach of All Men, Without Labour, by Powers of Nature and Machinery: An Address to All Intelligent Men, in Two Parts*. London: John Brooks, 1836 (reprint Nabu Press, 2010).

European Parliamentary Temporary Committee on the Echelon Interception System. "Report on the Existence of a Global System for the Interception of Private and Commercial Communications (ECHELON Interception System)." http://www.europarl.europa.eu/sides/getDoc.do?pubRef=-//EP//TEXT+REPORT+A5-2001-0264+0+NOT+XML+V0//EN&language=EN.

Farber, Erich. "Solar Water Heating and Space Heating in Florida." *Solar Energy* 3, no. 3 (October 1959): 21–25.

Felker, Fort. Telephone interview. March 3, 2010.

Fetzer, William. Personal interview.

Finney, John. "Nuclear-Power Costs Reduced as Atomic Fuels Exceed Their Life Expectancy." *New York Times*. June 14, 1964.

Fishman, Bob. Telephone interview. April 21, 2008.

Flint, Charles Louis. *American Farming and Stock Raising*. New York: Casselberry Company, 1892.

Focht, Jack. "An Ecosystem Is a Partnership in Nature." *New York Times*. February 25, 1968.

Foran, Jack. "The Day They Turned the Falls On: The Invention of the Universal Electrical Power System." University at Buffalo Libraries. http://ublib.buffalo.edu/libraries/projects/cases/niagara.htm.

Frank, D. J. "Science, Nature, and the Globalization of the Environment: 1870-1990." *Social Forces* 76, no. 2 (December 1997): 409–435.

Galbraith, John Kenneth. *The Affluent Society*. New York: Houghton Mifflin, 1976.

Garber, Kent, "John Doerr: U.S. Needs More Energy Research." *U.S. News and World Report*. October 22, 2009. http://politics.usnews.com/news/best-leaders/articles/2009/10/22/john-doerr-us-needs-more-energy-research.html.

Garud, Raghu and Peter Karnoe. "Bricolage Versus Breakthrough: Distributed and Embedded Agency in Technology Entrepreneurship." *Research Policy* 32, no. 2 (February 2003): 277–300.

General Compression. "General Compression Closes on $17 Million of Financing Commitments To Build Utility-Scale Energy Storage System." February, 23, 2010. http://www.generalcompression.com/pressroom/GC%20A%20Round%20Press%20Release%2023Feb10.pdf.

"G. H. Putnam Left Estate of $675,540." *New York Times*. April 21, 1932.

Gilinsky, Victor. "Nuclear Safety Regulation Lessons from US Experience." *Energy Policy* 20, no. 8 (August 1992): 704–711.

Gipe, Paul. "Design as if People Matter: Aesthetic Guidelines for a Wind Power Future." In *Wind Power in View: Energy Landscapes in a Crowded World*, ed. Martin J. Pasqualetti, Paul Gipe, and Robert W. Righter, 173–210. New York: Academic Press, 2002.

———. "Wind Energy Comes of Age: California and Denmark." *Energy Policy* 19, no. 8 (October 1991): 756–767.

———. *Wind Power: Renewable Energy for Home Farm and Business*. White River Junction, VT: Chelsea Green Publishing, 1993.

Goetzberger, Adolf et al. "Solar Cells: Past, Present, and Future." *Solar Energy Materials & Solar Cells* 74 (2002): 1–11.

Goldman, Arnold. *A Working Paper on Project Luz*. Unpublished manuscript.

———. Personal interview. Oakland, CA. June 26, 2009.

Google. "Google's Goal: Renewable Energy Cheaper than Coal." Press Release. November 27, 2007. http://www.google.com/intl/en/press/pressrel/20071127_green.html.

Gray, Thorne. "Tax Break for Solar Company Gains." *Sacramento Bee*. January 17, 1991.

Green, Joshua. "The Elusive Green Economy." *The Atlantic*. July 2009. http://www.theatlantic.com/magazine/archive/2009/07/the-elusive-green-economy/7554/.

Greenblatt, Jeffery B., Samir Succar, David C. Denkenberger, Robert H. Williams, Robert H. Socolow. "Baseload Wind Energy: Modeling the Competition Between Gas Turbines and Compressed Air Energy Storage for Supplemental Generation." *Energy Policy* 35, no. 3 (March 2007): 1474–1492.

Gross, Bill. Telephone interview. March 1, 2010.

"Growth of Idea Is Told by Inventor: Factory Forced to Move to Larger Quarters Twice as Demand Grows." *Los Angeles Times*. January 1, 1923.

Gutowski, Timothy et al. "Environmental Life Style Analysis (ELSA)." IEEE International Symposium on Electronics and the Environment. May 19–20, 2008. San Francisco, CA. http://web.mit.edu/ebm/www/Publications/ELSA%20IEEE%202008.pdf.

Hallacy, Daniel S. Jr. *The Coming Age of Solar Energy*. New York: Harper & Row, 1973.

Hammond, Allen, and William Metz. "Solar Energy Research: Making Solar After the Nuclear Model?" *Science* 197. July 15, 1977, 241.

Hansen, Zeynep, and Gary Libecap. "Small Farms, Externalities, and the Dust Bowl of the 1930s." *Journal of Political Economy* 112, no. 3 (2004): 665–694.

Hardin, Garrett. "The Tragedy of the Commons." *Science* 162, no. 859 (December 13, 1968): 1243–1248.

"Harnessing the Wind." *Time*. September 8, 1941. www.time.com/time/magazine/article/0,9171,849476,00.html.

"Harold R. Hay: Solar Pioneer." *Mother Earth News*. September/October 1976. www.motherearthnews.com/Renewable-Energy/1976-09-01/Passive-Cooling-Expert-Harold-Hay.aspx.

Harper, Peter. "Transfiguration among the Windmills." *Undercurrents* 5 (Winter 1973): 7–17.

Harris, Paul G., ed. *Climate Change and American Foreign Policy*. New York: St. Martin's Press, 2000.

Haukos, David A. "Fish and Wildlife Leaflet." Fish and Wildlife Service, 1992.

Hayes, Denis. "Washington Decrees a Solar Eclipse." *The New York Times*. August 12, 1981.

———. Telephone interview. January 8, 2010.

Hiester, T. R., and W. T. Pennell. "The Meterological Aspects of Siting Large Wind Turbines." January 1981. Technical report, PNL-2522. Pacific Northwest Laboratory. Richland, WA, 219–222.

Hirsh, Richard F. "PURPA: The Spur to Competition and Utility Restructuring." *The Electricity Journal* 12, no. 7 (August 9, 1999): 60–72.

———. *Technology and Transformation in the American Electric Utility Industry. Change in the Utility Industry*. Cambridge: Cambridge University Press, 2003.

Historic American Engineering Record. "Death Valley Ranch, Solar Heater, Death Valley Junction Vicinity, Inyo County, CA." Library of Congress. http://hdl.loc.gov/loc.pnp/hhh.ca1703.

Hoen, Ben, Ryan Wiser, Peter Cappers, Mark Thayer, and Gautam Sethi. "The Impact of Wind Power Projects on Residential Property Values." Lawrence Berkeley National Laboratory. December 2009. http://eetd.lbl.gov/ea/ems/reports/lbnl-2829e.pdf.

Hong, Tianzhen, S. K. Chou, and T. Y. Bong. "Building Simulation: An Overview of Developments and Information Sources." *Building and Environment* 35, no. 4 (May 1, 2000): 347–361.

Huet, Pierre-Daniel. *Memoirs of the Life of Peter Daniel Huet, Bishop of Avranches, Volume 2*. London: Longman, Hurst, Rees, and Orme, 1810.

Hughes, Thomas P. *Human-Built World: How to Think about Technology and Culture*. Chicago: University of Chicago Press, 2004.

———. "Technological Momentum in History: Hydrogenation in Germany: 1898–1933." *Past and Present*, no. 44 (August 1969): 106–132.

Hugill, Peter. "Good Roads and the Automobile in the United States 1880–1929." *Geographical Review* 72, no. 3 (July 1982): 327–349.

Hunter, Louis C. *Waterpower in the Century of the Steam Engine*. Charlottesville: Eleutherian Mills-Hagley Foundation, University Press of Virginia, 1979.

Hutchinson, F. W. "The Solar House." *Heating and Ventilating* 44 (March 1947): 55–59.

Hyman, Leonard S. *America's Electric Utilities: Past, Present, and Future*. Arlington, VA: Public Utilities Reports, 1988.

Ibenholt, Karin, "Explaining Learning Curves for Wind Power." *Energy Policy* 30, no. 13 (October 2002): 1181–1189.

Isenstadt, Sandy. "Visions of Plenty: Refrigerators in America around 1950." *Journal of Design History* 11, no. 4 (1998): 311–321.

Jacobs, Michael E. "Denis Hayes Is Named Director of Solar Energy Institute." *Physics Today*. September 1979, 105.

Jasper, James M. *The Art of Moral Protest: Culture, Biography and Creativity in Social Movements*. Chicago: University of Chicago Press, 1998.

Jasper, James M., and Jane D. Poulsen. "Recruiting Strangers and Friends: Moral Shocks and Social Networks in Animal Rights and Anti-Nuclear Protests." *Social Problems* 42, no. 4 (November 1995): 493–512.

Johansen, Jeff. Personal interview. December 13, 2009.

"John Doerr Sees Salvation and Profit in Greentech." TED.com. May 2007. http://www.ted.com/talks/lang/eng/john_doerr_sees_salvation_and _profit_in_greentech.html.

Johnson, John W. *Insuring Against Disaster*. Macon, GA: Mercer University Press, 1986.

Johnson, Keith. "Solar Power: Siemens, Solel, and the Sahara Desert." Environmental Capital, *Wall Street Journal*. October 16, 2009. http://blogs.wsj.com/environmentalcapital/2009/10/16/solar-power-siemens-solel-and-the-sahara-desert/tab/article/.

Johnson, Lyndon B. "Commencement Address at Holy Cross College." http://www.presidency.ucsb.edu/ws/index.php?pid=26305.

Junfeng, Li, and Hu Runqing. "Solar Thermal in China: Overview and Perspectives of the Chinese Solar Thermal Market." *Refocus* 6, no. 5 (September–October 2005): 25–27.

Kaika, Maria. "Dams as Symbols of Modernization: The Urbanization of Nature Between Geographical Imagination and Materiality." *Annals of the Association of American Geographers* 96, no. 2 (June 2006): 276–301.

Kammen, Dan. December 2009 Google San Francisco event.

Kamp, Linda. M., E. H. M. Ruud, and Cornelis D. Andriesse. "Notions on Learning Applied to Wind Turbine Development in the Netherlands and Denmark." *Energy Policy* 32 (2004): 1625–1637.

Kanellos, Michael. "Algae Startup #57: Is Heat the Secret to Success in Algae Fuel?" Greentechmedia.com. http://www.greentechmedia.com/green-light/post/is-heat-the-secret-to-success-in-algae-fuel/.

———. "With Greenfuel's Demise, What Happens to Algae?" Greentechmedia.com. May 13, 2009. http://www.greentechmedia.com/green-light/post/will-greenfuels-demise-what-happens-to-algae-4673/.

Kang, Byoungwoo and Gerbrand Ceder. "Battery Materials for Ultrafast Charging and Discharging." *Nature* 458 (March 12, 2009): 190–193.

Kapstein, Ethan. "Transition to Solar: An Historical Approach." In *Energy Transitions: Long-Term Perspectives*. Washington DC: Westview Press for the American Association for the Advancement of Science, 1981.

Karnoe, Peter. "Technological Innovation and Industrial Organization in the Danish Wind Industry." *Entrepreneurship and Regional Development* 2, no. 2 (1990): 105–123.

Kazmerski, Lawrence L. "Solar Photovoltaics R&D at the Tipping Point: A 2005 Technology Overview." *Journal of Electron Spectroscopy and Related Phenomena* 150 (2006): 105–135.

Kearney, David W. "Parabolic Trough Collector Overview." Presentation at the Parabolic Trough Workshop at NREL, 2007, Golden, CO.

Keats, John. *The Crack in the Picture Window*. New York: Houghton Mifflin, 1957.

Kedrosky, Paul. "John Doerr's Tears." PaulKedrosky.com. March 9, 2007. http://paul.kedrosky.com/archives/2007/03/09/the_case_of_vc.html.

Kennedy, Alexander Blackie William. *Compressed Air: Experiments upon the Transmission of Power by Compressed Air*. New York: D. Van Nostrand Company, 1892.

Kingsberry, Richard. Interviewed by Kathy Tucker. February 3, 2000. Hosted by the Center for Columbia River History.

Kintisch, Eli. *Hack the Planet*. Hoboken, NJ: John Wiley and Sons, 2010.

Kirk, Andrew. *Counterculture Green: The Whole Earth Catalog and American Environmentalism*. Lawrence: University of Kansas Press, 2007.

Kirsch, David. *The Electric Vehicle and the Burden of History*. New Brunswick, NJ: Rutgers University Press, 2000.

Kirsch, David, and Gijs Mom. "Visions of Transportation." *Business History Review* 76, no. 1 (Spring 2002): 75–110.

Kirsch, Jonathan. "Book Review: The Utopia of Democracy in Miniature." *Los Angeles Times*. September 27, 1989.

Knopp, Lisa. "Mammoth Bones." *Interdisciplinary Studies in Literature and the Environment* 9, no. 1 (Winter 2002): 189–202.

Knoshaug, Eric. *Current Status of the Department of Energy's Aquatic Species Program Lipid-Focused Algae Collection*. Conference poster presentation to the 31st Symposium on Biotechnology for Fuels and Chemicals. National Renewable Energy Laboratory, May 2009. http://www.nrel.gov/docs/fy09osti/45788.pdf.

Kotulak, Ronald. "Scientists Tell of Hopes for New Atom Plant." *Chicago Tribune*. June 11, 1964.

Kraemer, Susan. "California Gets Smart-Grid Funds to Bottle Wind." Clean Technica. November 26, 2009. http://cleantechnica.com/2009/11/26/california-gets-smart-grid-funds-to-bottle-wind/.

Kristensen, Leif, "Fragments of the Cup Anemometer History." February 14, 2005. http://www.cup-anemometer.dk/technical/The%20Cup%20Anemometer%20History.pdf.

Kryza, Frank T. *The Power of Light: The Epic Story of Man's Quest to Harness the Sun*. New York: McGraw-Hill, 2003.

Kubiszewski, Ida. "Odeillo Font-Romeau, France." The Encyclopedia of
Earth. http://www.eoearth.org/article/Odeillo_Font-Romeu,_France.

Kutleb, Charles. "Can Forests Bring Rain?" *Forest History* 15, no. 3
(October 1971): 14–21.

Laird, Frank. "Constructing the Future: Advocating Energy Technologies in
the Cold War." *Technology and Culture* 44, no. 1 (January 2003): 27–49.

——. *Solar Energy, Technology Policy, and Institutional Values.*
Cambridge: Cambridge University Press, 2001.

LaMonica, Martin. "Green-tech Venture Investing Cools Off in 2009." cnet
News. December 30, 2009. http://news.cnet.com/8301-11128_3-
10423018-54.html.

——. "Q&A: eSolar Bets on Software to Make Solar Cheaper." cnet News.
August 4, 2009. http://news.cnet.com/8301-11128_3-10302893-54.html.

Lee, Kaiman, and Linda Donnelly. *Solar Failure.* Newtonville, MA:
Environmental Design and Research Center, 1980.

Lesser, Jonathan A., and Xuejuan Su. "Design of an Economically Efficient
Feed-in Tariff Structure for Renewable Energy Development." *Energy
Policy* 36, no. 3 (March 2008): 981–990.

Levy, Harriett Lane. *920 O'Farrell Street: A Jewish Girlhood in San
Francisco.* Berkeley, CA: Heyday Press, 1996.

Lewis, Paul. *The Romance of Water-power.* London: S. Low, Marston, and
Co., 1931.

Lifsher, Marc. "You Can Thank Arthur Rosenfeld for Energy Savings." *Los
Angeles Times.* January 11, 2010.

Linstromberg, Robin and James Ballowe. "Thoreau and Etzler: Alternative
Views of Economic Reform." *American Studies* 11, no.1 (Spring 1970):
20–29.

Lof, G. O. G., and R. A. Tybout. "Cost of House Heating with Solar
Energy." *Solar Energy* 14 (1973): 253–278.

Lotker, Michael. "Barriers to Commercialization of Large-Scale Solar
Electricity: Lessons Learned from the LUZ Experience." Sandia
National Laboratory Report: SAND-91-7014. November 1991.
http://www.nrel.gov/csp/troughnet/pdfs/sand91_7014.pdf.

——. Telephone interview. December 7, 2009.

Lovins, Amory. *Soft Energy Paths.* San Francisco: Friends of the Earth
International, 1977.

"Luz Casts A Long Shadow." *Sacramento Bee.* April 11, 1991.

Lyford, William Gilman. *The Western Address Directory.* Baltimore, MD: Jos. Robinson, 1837.

MacDonald, Allan. "Lowell: A Commercial Utopia." *The New England Quarterly* 10, no. 1 (March 1937): 37–62.

Macdonald, Gregor. "Earth Day and Coal: It's Hard to Win a Fight Against a Cheap BTU." Gregor.us. April 22, 2009. http://gregor.us/coal/earth-day-and-coal-its-hard-to-win-a-fight-against-a-cheap-btu/.

Mackay, Alexander. *The Western World, or, Travels in the United States in 1846–47.* Vol. 3. New York: Negro Universities Press, 1968.

Madrigal, Alexis. "DOE Report Says More Wind Than Coal Planned for US Grid." *Wired.* June 2, 2008. http://www.wired.com/wiredscience/2008/06/new-doe-report/.

Malone, Patrick. *Waterpower in Lowell.* Baltimore, MD: The Johns Hopkins University Press, 2009.

Marcus, David. Telephone interview. March 5, 2010.

Mattson, Kevin. *What the Heck Are You Up To, Mr. President?* New York: Bloomsbury, 2009.

McCarthy, Tom. *Auto Mania: Cars, Consumers, and the Environment.* New Haven, CT: Yale University Press, 2007.

McGovern, George. "Introduction." In *Defending the Environment: A Strategy for Citizen Action,* by Joseph L. Sax. New York: Knopf, 1971.

McVeigh, James, Dallas Burtraw, Joel Darmstadter, and Karen Palmer. "Winner, Loser, or Innocent Victim? Has Renewable Energy Performed as Expected?" Resources for the Future discussion paper, 99-28. Washington, DC, 1999.

Meadows, Donella. "Let's Have a Little More Feedback." The Donella Meadows Archive. http://www.sustainer.org/dhm_archive/search.php?display_article=vn311feedbacked.

Mehta, Ardeshir. "Kibbutz Life in Israel During the Late Sixties and Early- to Mid-Seventies." 2001. http://homepage.mac.com/ardeshir/KIBBUTZ-March%2701.html.

Meinel, Aden. Telephone interview. February 9, 2010.

Melville, George W. "The Engineer and the Problem of Aerial Navigation." *The North American Review* 173, no. 541 (December 1901): 820–831.

Miller-Rushing, Abraham. "Global Warming and Flowering Times in Thoreau's Concord." *Ecology* 89, no. 2 (2008): 332–341.

Mitchell, Garry. "Power Plant Uses Compressed Air to Make Energy." Associated Press. *The Tuscaloosa News*. September 23, 1991.

Moezzi, Mithra and Rick Diamond. "Is Efficiency Enough? Towards a New Framework for Carbon Savings in the California Residential Sector." Public Interest Research Program Report. Berkeley: Lawrence Berkeley National Laboratory, 2005. http://escholarship.org/uc/item/205043pk.

Mollan, Charles. "Stopping the Sun: Stay Still, Won't You?" *Europhysics News*. March/April 1998.

Mom, Gijs. *The Electric Vehicle: Technology and Expectations in the Automobile Age*. Baltimore, MD: Johns Hopkins University Press, 2004.

Montrie, Chad, "'I Think Less of the Factory Than of My Native Dell': Labor, Nature, and the Lowell 'Mill Girls'." *Environmental History* 9, no. 2 (April 2004): 275–295.

Moore, J. Glen. "The Role of Congress." In *Implementation of Solar Thermal Technology*, ed. Ronal W. Larson and Ronal Emmett West, 70–118. Cambridge, MA: MIT Press, 1996.

———. "Solar Energy and the Reagan Administration." The Library of Congress Congressional Research Service. Mini brief number MB81265. UNT Digital Library. http://digital.library.unt.edu/ark:/67531/metacrs8799/.

Morris, Henry and Pedro Salom. "Letter to Western Society of Engineers." *Journal of the Western Society of Engineers* 2, no. 5 (October 1897): 547–548. Chicago: Western Society of Engineers.

Mortensen, Bent Ole Gram. "International Experiences of Wind Energy." *Environmental and Energy Law and Policy Journal* 2, no. 2 (2008): 179–209.

Moszeter, Jean. Interviewed by Kathy Tucker. April 12, 2000. Hosted by the Center for Columbia River History.

Mouawad, Jad. "Exxon to Invest Millions to Make Fuel from Algae." *New York Times*. July 13, 2009.

Moutoux, Richard David. "Wind Integrated Compressed Air Storage in Colorado." Master's thesis, University of Colorado, 2007. http://www.colorado.edu/engineering/energystorage/files/Moutoux_Thesis.pdf.

Mrozowski, S. A., E. L. Bell, M. C. Beaudry, D. B. Landon, and G. K. Kelso. "Living on the Boott: Health and Well Being in a Boardinghouse Population." *World Archaeology* 21, no. 2 (October 1989): 298–319.

"Mr. William C. Whitney's New York House." *House Beautiful* 10, no. 3 (August 1901): 131–142. Chicago: Herbert S Stone & Company.

Mumford, Lewis. *The Pentagon of Power.* New York: Harcourt Brace Jovanovich, 1974.

Murphy, Edward Charles. *The Windmill: Its Efficiency and Economic Use.* U.S. Geological Survey Paper no. 41–42. Washington DC: Government Printing Office, 1901.

Murphy, Lawrence. "The Use of Solar Thermal Energy to Generate Electricity." July 1981. SERI/TP-632-1287. www.scribd.com/doc/17381862/The-Use-of-Solar-Thermal-Energy -to-Generate-Electricity-1981.

Murphy, Priscilla Coit. *What a Book Can Do: The Publication and Reception of Silent Spring.* Amherst: University of Massachusetts Press, 2007.

Naar, Jon. E-mail to the author. December 15, 2008.

Nakhamkin, Personal interview. March 5, 2010.

National Park Service. "Historic Structure Report: Death Valley Scotty Historic District Main House and Annex." Death Valley Ranch, Death Valley National Monument, California/Nevada, U.S. Dept. of the Interior, 1991.

———. "Scotty's Castle—Behind the Scenes." nps.gov. www.nps.gov/deva/historyculture/scottys-bts-pag11.htm.

National Renewal Energy Laboratory. "Photovotalic Visualization." http://openpv.nrel.gov/visualization/index.

National Research Council, Solar Energy Research Committee. *Establishment of a Solar Energy Research Institute.* Washington, DC: National Academy of Sciences, 1975.

Navarro, James. "Green Scene: Solar Power Dreamin'." Defenders of Wildlife. Fall 2009. www.defenders.org/newsroom/defenders _magazine/fall_2009/green_scene_solar_power_dreamin.php.

Nemet, Gregory. "Beyond the Learning Curve: Factors Influencing Cost Reductions in Photovoltaics." *Energy Policy* 34, no. 17 (2006): 3218–3232.

———. "Demand-Pull Energy Technology Policies, Diffusion and Improvements in California Wind Power." In *Innovation for a Low Carbon Economy: Economic, Institutional and Management*

Approaches, ed. Timothy J. Foxon, Jonathan Kohler, and Christine Oughton, 47–78. Northampton, MA: Edward Elgar Publishing, 2008.

———. Telephone interview. March 15, 2010.

"New Power for Elevated." *New York Times*, February 2, 1899.

Nierenberg, R. "Free-Flow Variability on the Jess and Souza Ranches, Altamont Pass." Department of Energy technical report. March 1989.

Norcliffe, Glen. "Popeism and Fordism: Examining the Roots of Mass Production." *Regional Studies* 31, no. 3 (1997): 267–280.

Nordhaus, Ted, and Michael Shellenberger. *Break Through: From the Death of Environmentalism to the Politics of Possibility*. New York: Houghton Mifflin, 2007.

"Notes by a Roving Contributor—13." *Machinery*. August, 1899.

Noyes, John Humphrey. *History of American Socialisms*. Philadelphia, PA: J. P. Lippincott, 1870.

Nye, David. *American Technological Sublime*. Cambridge, MA: MIT Press, 1994.

———. *Consuming Power*. Cambridge, MA: MIT Press, 1999.

———. *Electrifying America: Social Meanings of a New Technology, 1880–1940*. Cambridge, MA: MIT Press, 1992.

Obama, Barack. "Remarks by the President at the National Academy of Sciences Annual Meeting." The White House. April 27, 2009. http://www.whitehouse.gov/the_press_office/Remarks-by-the-President-at-the-National-Academy-of-Sciences-Annual-Meeting/.

———. "Remarks by the President Challenging Americans to Lead the Global Economy in Clean Energy: Massachusetts Institute of Technology, Boston, Massachusetts." The White House. October 23, 2009. http://www.whitehouse.gov/the-press-office/remarks-president-challenging-americans-lead-global-economy-clean-energy.

Odell, Rice. *Environmental Awakening: The New Revolution to Protect the Earth*. Cambridge: Ballinger Publishing Co, 1980.

Odum, Howard. *Environment, Power, and Society*. New York: Wiley-Interscience, 1971.

Ogden, Peter, John Podesta, and John Deutch. "A New Strategy to Spur Energy Innovation." *Issues in Science and Technology*. Winter 2008. http://www.issues.org/24.2/ogden.html.

Pacific Northwest National Laboratory. "Meteorological Siting." *National Wind Atlas*. http://rredc.nrel.gov/wind/pubs/atlas/.

Paparian, Michael, "Solar Power." Letter to the editor. *Sacramento Bee*.

"Parabolic Trough Solar Field Technology." National Renewable Energy Laboratory. http://www.nrel.gov/csp/troughnet/solar_field.html.

Parisi, Anthony J. "New Path for Solar Institute." *New York Times*. December 26, 1979.

Parrish, Michael. "How Sun Failed to Shine on Solar Firm's Dreams." *Los Angeles Times*. December 8, 1991, 1.

Paulding, James Kirke. *The New Mirror for Travelers and Guide to the Springs*. New York: G. & C. Carvill, 1828.

"P. C. Putnam Bankrupt." *New York Times*. October 28, 1934.

Perelman, Lewis J. "Speculations on the Transition to Sustainable Energy." *Ethics* 90 (April 1980): 392–416.

Perlin, John. *From Space to Earth: The Story of Solar Electricity*. Ann Arbor, MI: Aatec Publications, 1999.

Perry, Thomas. *Experiments with Windmills*. U.S. Geological Survey Paper, no.20. Washington DC: Government Printing Office.

"Pertinent Facts of the Kenetech Fiasco." *Windpower Monthly*. July 1, 1997. Available behind paywall at http://www.windpowermonthly.com.

Pickard, William F., Amy Q. Shen, and Nicholas J. Hansing. "Parking the Power: Strategies and Physical Limitations for Bulk Energy Storage in Supply-Demand Matching on a Grid Whose Input Power Is Provided by Intermittent Sources." *Renewable and Sustainable Energy Reviews* 13, no. 8 (October 2009): 1934–1945.

Pimental, David et al. "Renewable Energy: Current and Potential Issues." *BioScience* 52, no. 12 (December 2002): 1111–1120.

"Pneumatic Clocks." *New York Times*. August 14, 1881.

"Pneumatic Clocks." *Scientific American* XLIII, no. 2 (July 10, 1880), 19.

Pope, Albert A. "Export of American Bicycles." Official proceedings of the International Commercial Congress. Philadelphia, PA: The Philadelphia Commercial Museum, 1899.

Power-Electronic, Variable-Speed Wind Turbine Development: 1988–1993. Electric Power Research Institute technical report TR-104738. November 16, 1995.

President's Materials Policy Commission. *Resources for Freedom*. Washington, DC: Government Printing Office, 1952.

Price, Hank. "Testimony to the Public Utilities Commission of Colorado, 2007." Interwest Energy Alliance. http://www.interwestenergy.org/documents/documents/2008-04-28_07A-447_Price_Answer_Testimony.pdf.

Prospectus for the Wave Power and Air Compressing Company, 1895. California Historical Society.

"Purdue Experiments Show How Sun's Rays Can Help to Reduce Heating Cost in Home." *New York Times*. December 1, 1946.

Pursell, Carroll. "The Rise and Fall of the Appropriate Technology Movement in the United States, 1965–1985." *Technology and Culture* 34, no. 3 (July 1993): 629–637.

Putnam, Palmer Cosslett. "A Reconnaissance Among Some Volcanoes of Central America." Master's Thesis, MIT, 1924.

———. *Power from the Wind*. Van Nostrand: New York, 1948.

Putnam, R. D. "The Strange Disappearance of Civic America." *The American Prospect*. December 1, 1995. http://www.prospect.org/cs/articles?articleId=4972

Rae, John B. "The Electric Vehicle Company: A Monopoly that Missed." *The Business History Review* 29, no. 4 (December 1955): 298–311.

Rashad, S. M., and F. H. Hammad. "Nuclear Power and the Environment: Comparative Assessment of Environmental and Health Impacts of Electricity-Generating Systems." *Applied Energy* 65, nos. 1–4 (April 2000): 211–229.

"Reddy Kilowatt Commercial." YouTube. Posted August 4, 2007. www.youtube.com/watch?v=PznxZ3zmL00.

"Redondo Pier Sinks in Sea; Starr Wave Motor Follows It to Bottom." *Los Angeles Times*. February 13, 1909.

Reece, Ray. *The Sun Betrayed: A Report on the Corporate Seizure of U.S. Solar Energy Development*. Boston: South End Press, 1979.

Reich, Charles. *The Greening of America: How the Youth Revolution Is Trying to Make America Livable*. New York: Random House, 1970.

Reichman, Edward. "Re: Luz." E-mail to the author. December 6, 2009.

Reichman, Edward, and Fred Rosner. "The Bone Called Luz." *Journal of the History of Medicine and Allied Sciences* 51, no. 1 (1996): 52–65.

Renewable Energy Policy Network for the 21st Century. "Renewables Global Status Report 2009." ren21.net. www.ren21.net/globalstatusreport/g2009.asp.

Reynolds, Alva. "Pile Cleaner and Protector." U.S. Patent No. 1266051, 1918. (See http://www.google.com/patents?id=eyRAAAAAEBAJ&printsec =abstract&zoom=4&source=gbs_overview_r&cad=0#v=onepage &q&f=false.)

"Reynolds Wave Motor and the Inventors." *Los Angeles Herald.* March 17, 1907.

Richardson, James. "Energy Firm Pushing Hard for Property Tax Exemption." *Sacramento Bee.* March 20, 1991.

———. "Tax Break Quickly Ok'd for Solar Energy Plant—Bill Assailed as 'Giveaway' Goes to Wilson." *Sacramento Bee.* April 9, 1991.

Righter, Robert. *Wind Energy in America: A History.* Norman: University of Oklahoma Press, 1996.

Robert Dennis Collection of Stereoscopic Views. New York Public Library. http://digital.nypl.org/dennis/stereoviews/browse.html.

Rome, Adam. *The Bulldozer in the Countryside: Suburban Sprawl and the Rise of American Environmentalism.* Cambridge: Cambridge University Press, 2001.

———. "Give Earth a Chance." *The Journal of American History* 90, no. 2 (September 2003): 525–555.

———. Telephone interview. March 18, 2010.

Romm, Joe. "The Technology That Will Save Humanity." *Salon.* April 14, 2008. www.salon.com/news/feature/2008/04/14/solar_electric _thermal.

Rose, Anne C. *Transcendentalism as a Social Movement: 1830–1850.* New Haven, CT: Yale University Press, 1981.

Rosenberg, Nathan, and Manuel Trajtenberg. "A General-Purpose Technology at Work: The Corliss Steam Engine in the Late-Nineteenth-Century United States." *The Journal of Economic History* 64 (2004): 61–99.

Rosenfeld, Art. "The Art of Energy Efficiency: Protecting the Environment with Better Technology." *Annual Review of Energy and the Environment* 24 (1999): 33–82.

———. "A Brave New Source: Physicists Discover Energy Efficiency." Talk delivered October 14, 2008 at UC-Davis. California Energy Commission. www.energy.ca.gov/commissioners/videos/Rosenfeld_ Main.php?movieType=MOV.

———. "Dr. Rosenfeld Video." Talk delivered November 4, 2005 at Portland State University. California Energy Commission. www .energy.ca.gov/commissioners/videos/Rosenfeld_Main.php ?movieType=MOV.

Rosenfeld, Arthur H., with Deborah Poskanzer. "A Graph Is Worth a Thousand Gigawatt-Hours: How California Came to Lead the United States in Energy Efficiency." *Innovations: Technology, Governance, Globalization* 4, no. 4 (Fall 2009): 57–79.

Rowlands, Ian H. "Envisaging Feed-In Tariffs for Solar Photovoltaic Electricity: European Lessons for Canada." *Renewable and Sustainable Energy Reviews* 9, no. 1 (February 2005): 51–68.

Rutledge, David. "Hubbert's Peak, the Coal Question, and Climate Change." Rutledge. http://rutledge.caltech.edu.

Ryan, Hugh McLaren. *High Voltage Engineering and Testing.* London: The Institution of Electrical Engineers, 2001.

Salom, Pedro G. "Automobile Vehicles." *Journal of the Franklin Institute* 141, nos. 841–846 (January–June 1896): 278–296.

Sanderson, Edgar, and John McGovern. *The World's History and Its Makers.* Vol. 10. Chicago: Universal History Publishing Company, 1900.

San Diego Renewable Energy Study Group. "Potential for Renewable Energy in the San Diego Region: Appendix E: Solar Thermal— Concentrated Solar Power." San Diego Regional Renewable Energy Study Group. August 2005. http://*www.renewablesg.org/docs/Web/ AppendixE.pdf.*

Sass, Stephen. *The Substance of Civilization: Materials and Human History from the Stone Age to the Age of Silicon.* New York: Arcade Publishers, 1998.

Saunders, W. L. *Compressed Air Information; or, A Cyclopedia Containing Practical Papers on the Production, Transmission and Use of Compressed Air.* New York: Compressed Air, 1903.

Savino, Joseph. "A Brief Summary of the Attempts to Develop Large Wind-Electric Generating Systems in the U.S." NASA Technical Memorandum, 1974.

Schatzberg, Eric. "The Mechanization of Urban Transit in the United States: Electricity and Its Competitors." In *Technological Competitiveness: Contemporary and Historical Perspectives on*

Electrical, Electronics, and Computer Industries, edited by William
Aspray, 225–242 (Piscataway, NJ: IEEE Press, 1993).

Schell, Orville. *The Town That Fought to Save Itself.* Film. New York:
Pantheon, 1976.

Schiff, Martin. "Neo-Transcendentalism in the New Left Counter-Culture:
A Vision of the Future Looking Back." *Comparative Studies in Society
and History* 15, no. 2 (March 1973): 130–142.

Schoenfeld, A. Clay, Robert F. Meier, and Robert J. Griffin. "Constructing
a Social Problem: The Press and the Environment." *Social Problems* 27,
no. 1 (October 1979): 38–61.

Schumacher, E. F. *Small Is Beautiful: Economics As If People Mattered.*
London: Blond and Briggs, 1973.

Schurr, Sam H, ed. *Energy, Economic Growth and the Environment.*
Baltimore, MD: Resources for the Future, 1972.

Seaborg, Glenn. *Peaceful Uses of Nuclear Energy.* Oak Ridge, TN: U.S.
Atomic Energy Commission, 1970.

"Secretary Chu Announcement of $151 Million in ARPA-E Grants." U.S.
Department of Energy. October 26, 2009.
http://www.energy.gov/news/8211.htm.

"Sees Broad Demand for the 'Solar' Home." *New York Times.* April 25,
1945.

Senate Committee on Public Works. *The Impact of Growth on the
Environment.* 93rd U.S. Cong., 1st session. 1973.

SERI: The First Year. Scrib'd. http://www.scribd.com/doc/17381198/SERI-
The-First-Year-1978.

Shapiro, Carl, and Hal R. Varian. "The Art of Standard Wars." *California
Management Review* 41, no. 2 (1999): 8–32.

"Share Price Volatile After Sell Direction Shakes Investors." *Wind Power
Monthly.* September 1994. Available behind paywall at
http://www.windpowermonthly.com/.

Sheehan, John, Terri Dunahay, John Benemann, and Paul Roessler. *A Look
Back at the U.S. Department of Energy's Aquatic Species Program:
Biodiesel from Algae.* National Renewable Energy Laboratory, July
1998. http://www.nrel.gov/docs/legosti/fy98/24190.pdf.

Shulman, Peter. Personal communication. December 11, 2008.

Simon, Maron, ed. *Your Solar House.* New York: Simon and Schuster, 1947.

Simons, Howard. "Dispute Flares Over A-Power Cost Dip." *Washington Post*. September 23, 1964.

Sinclair, Bruce. *Philadelphia's Philosopher Mechanics*. Baltimore, MD: Johns Hopkins University Press, 1974.

Skurka, Norma, and Jon Naar. *Design for a Limited Planet*. New York: Ballantine Books, 1976.

"Smarting from the Wind." *The Economist*. January 26, 2010. http://www.economist.com/node/15387364?story_id=15387364.

Smeaton, John. *Experimental Enquiry Concerning the Natural Powers of Wind and Water to Mills and Other Machines*. London: Printed for I. and J. Taylor, no. 56, High-Holborn, 1796.

Smil, Vaclav. "Moore's Curse and the Great Energy Delusion." *The American* 2, no. 6 (2008): 34–41.

———. *Energy at the Crossroads: Global Perspectives and Uncertainties*. Cambridge, MA: MIT Press, 2003.

Smith, D. R. "The Wind Farms of Altamont Pass." *Annual Review of Energy* 12 (1987): 145–183.

Smith, Henry Nash. "Rain Follows the Plow." *The Huntington Library Quarterly* 10, no. 2 (February 1947): 169–193.

Socolow, Robert H. "Reflections on the 1974 APS Energy Study." *Physics Today* 39, no. 1 (January 1986): 60–68.

"Solar Activist, Denis Hayes, Heads SERI." *Science* 205, August 10, 1979, 563.

Solar Energy Industries Association. "Utility-Scale Solar Projects in the United States." November 5, 2010. www.seia.org/galleries/pdf/Major%20Solar%20Projects.pdf.

"Solar Energy Method Found?" *Lawrence Daily Journal World*. February 3, 1972.

Solar Energy Research Institute. *A New Prosperity: Building a Sustainable Energy Future, The SERI Solar/Conservation Study*. Andover, MA: Brick House Publishing, 1981.

———. *Institutional Plan: Fiscal Years 1981– 1986*. Golden, CO: Solar Energy Research Institute, 1990.

"'Solar House' Owner Starts Homes Project." *Chicago Daily Tribune*. July 20, 1941.

"Solar Houses Win Approval Across Nation." *Washington Post*. October 17, 1948, R2.

Sorrell, Steve. "Jevons' Paradox Revisited: The Evidence for Backfire from Improved Energy Efficiency." *Energy Policy* 37, no. 4 (April 2009): 1456–1469.

Sorrell, Steve, and John Dimitropoulos and Matt Sommerville. "Empirical Estimates of the Direct Rebound Effect: A Review." *Energy Policy* 37 (2009): 1356–1371.

Spatari, S., M. Bertram, Robert B. Gordon, K. Henderson, and T. E. Graedel. "Twentieth Century Copper Stocks and Flows in North America: A Dynamic Analysis." *Ecological Economics* 54, no. 1 (July 1, 2005): 37–51.

Steckel, Richard. "Stature and Living Standards in the United States." In *American Economic Growth and Standards of Living before the Civil War*, ed. Robert E. Gallman and John Joseph Wallis, 265–310 (NBER Conference Report, Jul 20–22, 1990). Chicago: University of Chicago Press, 1992.

Steenson, Molly Wright. "A Regressive Glossary of the Poste Pneumatique: Tracing the Material of the Pneumatic Tube Network in Paris (1875–1890)." Unpublished manuscript.

Stegner, Wallace. *The American West as Living Space*. Ann Arbor: University of Michigan Press, 1987.

Stein, Matthew. *When Technology Fails: A Manual for Self-Reliance, Sustainability, and Surviving the Long Emergency*. White River Junction, VT: Chelsea Green Publishing, 2008.

Steinberg, Theodore. *Nature Incorporated: Industrialization and the Waters of New England*. Cambridge: Cambridge University Press, 2003.

Stewart, Homer J. Interviewed by John L. Greenberg. Archives California Institute of Technology. http://oralhistories.library.caltech.edu/135/.

Stoehr, Taylor. *Nay-Saying in Concord: Emerson, Alcott and Thoreau*. Hamden, CT: Archon Books, 1979.

Stoll, Steven. *The Great Delusion: A Mad Inventor, Death in the Tropics, and the Utopian Origins of Economic Growth*. New York: Hill and Wang, 2008.

Stoller, Leo. "A Note on Thoreau's Place in History of Phenology." *Isis* 47, no. 2 (June 1956): 172–181.

Stover, Dawn. "The Forecast for Wind Power." *Popular Science*. July 1995, 66–72, 85.

Stromberg, R. P. "A Status Report on the Sandia Laboratories Solar Total Energy Program." *Solar Energy* 17, no. 6 (1975): 359–366.

Sudarshan, Anant. "Deconstructing the Rosenfeld Curve." USAEE-IAEE
    Working Paper No. 10-057. December 6, 2010.
    http://ssrn.com/abstract=1715860.
"Sun Day Organizer Is New Solar Director." *Science News* 116, no. 5
    (August 4, 1979): 84.
"Sun Day: 'When EXXON Gets Control." *Ottawa Citizen*. May 5, 1978.
"Sun May Aid Heating of Postwar Homes." *Washington Post*. March 19,
    1944, M5.
Sylla, Richard. "The U.S. Banking System: Origin, Development, and
    Regulation." History Now.
    www.gilderlehrman.org/historynow/06_2010/historian2.php.
"Tax Law Generating Big Sun-Power Sales." *The Washington Post*. August
    27, 1981.
Thacker, Dean R., and Harold Babcock. "The Mass Culture of Algae."
    Solar Energy 1, no. 1 (January 1957): 37–50.
Thayer, Robert. "Altamont: Public Perceptions of a Wind Energy
    Landscape." *Landscape and Urban Planning* 14 (1987): 379–398.
"The Best Idea of All." *San Francisco Examiner*. May 5, 1895, 22, col. 2.
"The Compressed-Air Motor." *New York Times*, October 28, 1896.
"The Green Factor." *Time*. October 12, 1992.
"The President's Solar Advocate." *New York Times*. June 15, 1977.
The San Francisco Call. March 28, 1909, 59, image 59.
Thiele, Leslie Paul. *Environmentalism for a New Millennium: The
    Challenge of Coevolution*. Oxford and New York: Oxford University
    Press, 1999.
Thoreau, Henry David. *A Week on the Concord and Merrimack Rivers*.
    Boston: James R. Osgood and Company, 1873.
———. "Paradise (to be) Regained." *The United States Magazine and
    Democratic Review* 13 (November 1843): 451–463. New York:
    J. & H. G. Langley.
———. *Walden*. New York: Longmans, Green, and Co., 1910.
Torrey, Volta. *Wind Catchers: American Windmills of Yesterday and
    Tomorrow*. Brattleboro, VT: Stephen Greene Press, 1976.
Toth, Robert C. "U.S. A-Plant Estimates Challenged by Russian." *Los
    Angeles Times*. September 1, 1964.
Tuomi, Ilkka. "The Lives and the Death of Moore's Law." *First Monday*.
    November 2002. http://131.193.153.231/www/issues/issue7_11/tuomi/.

Turner, Fred. *From Counterculture to Cyberculture*. Chicago: University of Chicago Press, 2006.

Twain, Mark. *Connecticut Yankee in King Arthur's Court*. New York: Harper and Brothers, 1917.

"Untitled." Associated Press. May 4, 1978.

U.S. Congress, Office for Technology Assessment, *Energy Efficiency: Challenges and Opportunities for Electric Utilities*, OTA-E-561. Washington DC: U.S. Government Printing Office, 1993.

U.S. Congressional House Subcommittee on Energy and Power. *Permanent Extension of Certain Expiring Tax Provisions*. 102nd Cong., 2nd sess. Washington, DC: GPO, 1992.

U.S. Congressional Senate. *National Energy Policy Act of 1989*. Committee on Energy and Natural Resources. 101st Cong., 1st sess. Washington, DC: GPO, 1990.

U.S. Department of Agriculture. "World Agriculture Supply and Demand Estimates Report for March 2010." ISSN: 1554–9089. http://usda.mannlib.cornell.edu/usda/waob/wasde//2010s/2010/wasde-04-09-2010.txt.

U.S. Department of Energy. "GridWorks: Overview of the Electric Grid." http://sites.energetics.com/gridworks/grid.html.

Usher, Payson Abbott. *A History of Mechanical Inventions*. Mineola, NY: Dover Publications, 1988.

U.S. House of Representatives. "Economic Growth through Energy Efficiency." Hearing before the Subcommittee on Energy Conservation and Power of the Committee on Energy and Commerce. 97th Congress. May 1981.

UTEX. The Culture Collection of Algae at the University of Texas, Austin. http://web.biosci.utexas.edu/utex/protocols.aspx.

Vietor, Richard. *Energy Policy in America Since 1945*. Cambridge: Cambridge University Press, 1984.

Vigne, Godfrey Thomas. *Six Months in America*. Philadelphia, PA: Thomas T. Ash, 1833.

Wald, Matthew. "A New Era for Windmill Power." *New York Times*. September 8, 1992.

Waldau-Jager, Arnulf. "Status of Thin Film Solar Cells in Research, Production, and Market." *Solar Energy* 77 (2004): 667–678.

Walker, David Alan. "Biofuels, Facts, Fantasy, and Feasibility." *Journal of Applied Phycology* 21, no. 5 (2009): 509–517.

Walker, Jim. "Reliability and the Future of Wind Energy." Presentation at the 2005 Reliability Workshop, hosted by the DOE and Sandia National Laboratory. Livermore, California.

Wallace, Burton. "Unlimited Electric Power." *Overland Monthly* 50, no. 3 (September 1907): 289–291.

Walters, Dan. E-mail Message to author. November 19, 2009.

Wave Motor Patent Database. Inventing Green. http://www.greentechhistory.com/wave-power-patent-database/.

"Wave Power Chiefs Quit." *Los Angeles Times*, October 4, 1908.

Weart, Spencer. *The Discovery of Global Warming.* American Institute of Physicists. http://www.aip.org/history/climate/index.html.

Webb, Walter Prescott. *History as High Adventure.* Austin, TX: Pemberton Press, 1969.

———. *The Great Plains: A Study in Institutions and Environment.* Waltham, MA: Blaisdell Publishing Company, 1931.

Wei, Taoyuan. "A General Equilibrium View of Global Rebound Effects." Cicero Working Paper 2009:02. http://www.cicero.uio.no/media/7218.pdf.

Weil, Josh. "The Wind Farmers of East 11th Street." *New York Times.* August 3, 2008. http://www.nytimes.com/2008/08/03/nyregion/thecity/03wind.html?ref=thecity.

Weinberg, Alvin. "Can the Sun Replace Uranium?" Institute for Energy Analysis. 1977. Report, ORAU/IEA(M)-77-21.

———. *The First Nuclear Era: The Life and Times of a Technological Fixer.* New York: American Institute of Physics, 1994.

Weiss, Michael. "Everyone Loves Solar Energy, But. . . ." *New York Times Magazine.* September 24, 1989. http://www.nytimes.com/1989/09/24/magazine/everybody-loves-solar-energy-but.html?pagewanted=all.

Wellock, Thomas Raymond. *Critical Masses: Opposition to Nuclear Power in California, 1958–1978.* Madison: University of Wisconsin Press, 1998.

"Whatever Happened to Leadership and Where are the Leaders?" *Time* magazine advertisement, *New York Times.* July 8, 1974. 44.

White, Richard. *The Organic Machine: The Remaking of the Columbia River.* New York: Hill and Wang, 1995.

"Whitney Auto Smash: Chauffeur Kills a Peddler Near Paris and Is Himself Hurt." *New York Times.* August 21, 1907.

Whittier, John Greenleaf. "The Stranger in Lowell: Tales and Sketches." 1843. http://library.uml.edu/clh/All/jgwhi.htm.

"Who Owns It, Anyway?" *Life.* November 20, 1902.

Wicker, Tom. "Johnson Reports a 'Breakthrough' in Atomic Power." *New York Times.* June 11, 1964.

Wilber, Charles Dana. *The Great Valleys and Prairies of Nebraska and the Northwest.* Omaha, NE: Daily Republican Print, 1881.

Williams, Albert Nathaniel. *The Water and the Power.* New York: Duell, Sloan and Pearce, 1951.

Williams, James C. *Energy and the Making of Modern California.* Akron, OH: The University of Akron Press, 1997.

Williams, Selina. "U.K. Tries to Catch a Wave." *Wall Street Journal.* March 15, 2010.

Williamson, Jeffrey G. "Urban Disamenities, Dark Satanic Mills, and the British Standard of Living Debate." *The Journal of Economic History* 41, no. 1 (March 1981): 75–83.

Willis, Charles G., Brad Ruhfel, Richard B. Primack, Abraham J. Miller-Rushing, and Charles C. "Phylogenetic Patterns of Species Loss in Thoreau's Woods are Driven by Climate Change." *Proceedings of the National Academy of Sciences of the United States of America* 105, no. 44 (2008): 17029–17033.

"Wind Energy Resource Atlas." National Renewable Energy Laboratory. http://rredc.nrel.gov/wind/pubs/atlas/chp1.html.

Winner, Langdon. *The Whale and the Reactor: A Search for Limits in an Age of High Technology.* Chicago: University of Chicago Press, 1989.

Winston, R., and H. Hinterberger. "Principles of Cylindrical Concentrators for Solar Energy." *Solar Energy* 17, no. 4 (September 1975): 55–58.

Wolfe, Maynard Frank. *Rube Goldberg: Inventions.* New York: Simon & Schuster, 2000.

Wood, Daniel B. "State-of-the-Art Solar Collectors for Clean Energy–Luz System in Mojave Desert Boasts World's Most Advanced Solar-Thermal Technology." *Christian Science Monitor.* June 13, 1989, 12–13.

Woody, Todd. "Wind Jobs Outstrip Coal Mining." Green Wombat. January 29, 2009. http://thegreenwombat.com/2009/01/28/wind-jobs-outstrip-the-coal-industry/.

Wyman, Morril. "Practical Treatise on Ventilation." *American Journal of Medical Sciences* 18, no. 35 (July 1849): 129–47. London: Wiley & Putnam, and John Miller, 1849.

Yergin, Daniel. *The Prize: The Epic Quest for Oil, Money & Power.* New York: Free Press, 1991.

Yi, Matthew. "Tax Boost Predicted in Face of State's Largest Deficit Since '91." *San Francisco Chronicle.* December 13, 2007. http://articles.sfgate.com/2007-12-13/news/17274074_1_billion-budget-sales-taxes-tax-increase.

Young, Lori-Ann S. "The Paper Mill and the City of Camas." Student paper hosted at the Center for Columbia River History. www.ccrh.org/comm/camas/student%20papers/camas%20mill.htm.

Zahner, Robert. *The Transmission of Power by Compressed Air.* New York: D. Van Nostrand, 1878.

INDEX